Advancement in Microstrip Antennas with Recent Applications

Advancement in Microstrip Antennas with Recent Applications

Editor

Fredrik Jonasen

scitus academics

Advancement in Microstrip Antennas with Recent Applications
Edited by **Fredrik Jonasen**

ISBN: 978-1-68117-280-4
Library of Congress Control Number: 2016937452

www.scitusacademics.com
Box No. 4766, 616 Corporate Way,
Suite 2, Valley Cottage,
NY 10989

Printed in United States of America on Acid Free Paper

Preface

In telecommunication, there are several types of microstrip antennas the most common of which is the microstrip patch antenna or patch antenna. Microstrip patch antennas have become the favorite of antenna designers because of its versatility and advantages of planar profile, ease of fabrication, compatibility with integrated circuit technology, and conformability with a shaped surface. A patch antenna is a narrowband, wide-beam antenna fabricated by etching the antenna element pattern in metal trace bonded to an insulating dielectric substrate, such as a printed circuit board, with a continuous metal layer bonded to the opposite side of the substrate which forms a ground plane. A single patch antenna provides a maximum directive gain of around 6-9 dBi.Common microstrip antenna shapes are square, rectangular, circular and elliptical, but any continuous shape is possible. Some patch antennas do not use a dielectric substrate and instead are made of a metal patch mounted above a ground plane using dielectric spacers; the resulting structure is less rugged but has a wider bandwidth. Because such antennas have a very low profile, are mechanically rugged and can be shaped to conform to the curving skin of a vehicle, they are often mounted on the exterior of aircraft and spacecraft, or are incorporated into mobile radio communications devices. Microstrip antennas are relatively inexpensive to manufacture and design because of the simple 2-dimensional physical geometry. They are usually employed at UHF and higher frequencies because the size of the antenna is directly tied to the wavelength at the resonant frequency. The book, entitled Advancement in Microstrip Antennas with Recent Applications, discusses basic and advanced concepts of microstrip antennas, including design procedure and recent applications. It shall be of immense valuable tool for electrical and computer engineers and other scientists well versed in microstrip antenna technology.

Table of Contents

1

CHAPTER

DESIGN TECHNIQUES FOR CONFORMAL MICROSTRIP ANTENNAS AND THEIR ARRAYS

Daniel B. Ferreira[1], Cristiano B. de Paula[1] and Daniel C. Nascimento[2]

[1] CPqD - Telecommunications R&D Foundation, Brazil

[2] ITA - Technological Institute of Aeronautics, Brazil

1. INTRODUCTION

Owing to their electrical and mechanical attractive characteristics, conformal microstrip antennas and their arrays are suitable for installation in a wide variety of structures such as aircrafts, missiles, satellites, ships, vehicles, base stations, etc. Specifically, these radiators can become integrated with the structures where they are mounted on and, consequently, do not cause extra drag and are less visible to the human eye; moreover they are low-weight, easy to fabricate and can be integrated with microwave and millimetre-wave circuits [1,2]. Nonetheless, there are few algorithms available in the literature to assist their design. The purpose of this chapter is to present accurate design techniques for conformal microstrip antennas and arrays composed of these radiators that can bring, among other things, significant reductions in design time.

The development of efficient design techniques for conformal microstrip radiators, assisted by state-of-the-art computational electromagnetic tools, is desirable in order to establish clear procedures that bring about reductions in computational time, along with high accuracy results. Nowadays, the commercial availability of high performance three-dimensional electromagnetic tools allows computer-aided analysis and optimization that replace the design process based on iterative experimental modification of the initial prototype. Software such as CST®, which uses the Finite Integration Technique (FIT), and HFSS®, based on the Finite Element Method (FEM), are two examples of analysis tools available in the market [3]. But, since they are only capable of performing the analysis of the structures, the synthesis of an antenna needs to be guided by an algorithm whereby iterative process of simulations, result analysis and model's parameters modification are conducted until a set of goals is satisfied [4].

Generally, the design of a probe-fed microstrip antenna starts from an initial geometry determined by means of an approximate method such as the Transmission-line Model [5-7] or the Cavity Model [8]. Despite their numerical efficiency, i.e., they are not time-consuming and do not require a powerful computer to run on, these methods are not accurate enough for the design of probe-fed conformal microstrip antennas, leading to the need of antenna model optimization through the use of full-wave electromagnetic solvers in an iterative process. However, the full-wave simulations demand high computational efforts. Therefore, it is advantageous to have a design technique that employs full-wave electromagnetic solvers for accuracy purposes, but requires a small number of simulations to accomplish the design. Unfortunately, the approximated methods mentioned before provide no means for using the full-wave solution data in a feedback scheme, what precludes their integration in an iterative design process, hence restricting them just to the initial design step. In this chapter, in order to overcome this drawback and to reduce the number of full-wave simulations required to synthesize a probe-fed conformal microstrip antenna with quasi-rectangular patch, a circuital model able to predict the antenna impedance locus calculated in the full-wave electromagnetic solver is developed with the aim of replacing the full-wave simulations for the probe positioning. This is accomplished by the use of a transmission-line model with a set of parameters derived to fit its impedance locus to the one obtained in the full-wave simulation [4]. Since this transmission line model adapts its input impedance to fit the one from the full-wave simulation, at each algorithm iteration, it is an adaptive model per nature, so it was named ATLM – Adaptive Transmission Line Model. In Section 2, the ATLM is described in detail and some design examples are given to demonstrate its applicability.

Similar to what occurs with conformal microstrip antennas, the literature does not provide a great number of techniques to guide the design of conformal microstrip arrays. Among these design techniques, there are, for example, the Dolph-Chebyshev design and the Genetic Algorithms [9]. However, the results provided by the Dolph-Chebyshev design are not accurate for beam steering [10], once it does not take the radiation patterns of the array elements into account in its calculations, i.e., for this pattern synthesis technique, the array is composed of only isotropic radiators; hence it implies errors in the main beam position and sidelobes levels when the real patterns of the array elements are considered. On the other hand, the Genetic Algorithms can handle well the radiation patterns of the array elements and guarantee that the sidelobes assume a level better than a given specification *R* [9]. Nonetheless, to control the array directivity [11], it is important that all these sidelobes have the same level *R*, but to obtain this type of result Genetic Algorithms frequently requires a high number of iterations which increases the design time. Thus, in Section 3, an elegant procedure is employed, based on the solution of linearly constrained least squares problems [12], to the design of conformal microstrip arrays. Not only does this algorithm take the radiation pattern of each array element into account, but it also assures that a determined number of sidelobes levels have the same value, so to get optimized array directivity. And, to obtain more accurate results, the radiation patterns of the array elements, which feed the developed procedure, are evaluated from the array full-wave simulation data. In this work, the CST® Version 2012 was used to get these data. The proposed design technique was coded in the Mathematica® package [13] to create a computer program capable of assisting the design of conformal microstrip arrays. Some examples are given in this section to illustrate the use and effectiveness of this computer program.

Another concern for designing conformal microstrip arrays is how to implement a feed network that can impose appropriate excitations (amplitude and phase) on the array elements to synthesize a desired radiation pattern. Some microstrip arrays used in tracking systems, for example, employ the Butler Matrix [11] as a feed network. Nevertheless, this solution can just accomplish a limited set of look directions and cannot control the sidelobes levels. Hence, in this work, in order not to limit the number of radiation patterns that can be synthesized, an active circuit, composed of phase shifters and variable gain amplifiers, is adopted to feed the array elements. Expressions for calculating the phase shifts and the gains of these components are addressed in Section 4, as well as some design examples are provided to demonstrate their applicability.

2. ALGORITHM FOR CONFORMAL MICROSTRIP ANTENNAS DESIGN

The main property of the proposed ATLM is to allow the prediction of the impedance locus determined in the antenna full-wave analysis when one of its geometric parameters is modified, for instance, the probe position, thereby replacing full-wave simulations in probe position optimization. It results in a dramatic computational time saving, since a circuital simulation is usually at least 1000 times faster than a full-wave one. In this section, the ATLM is described in detail and some design examples are provided to highlight its advantages.

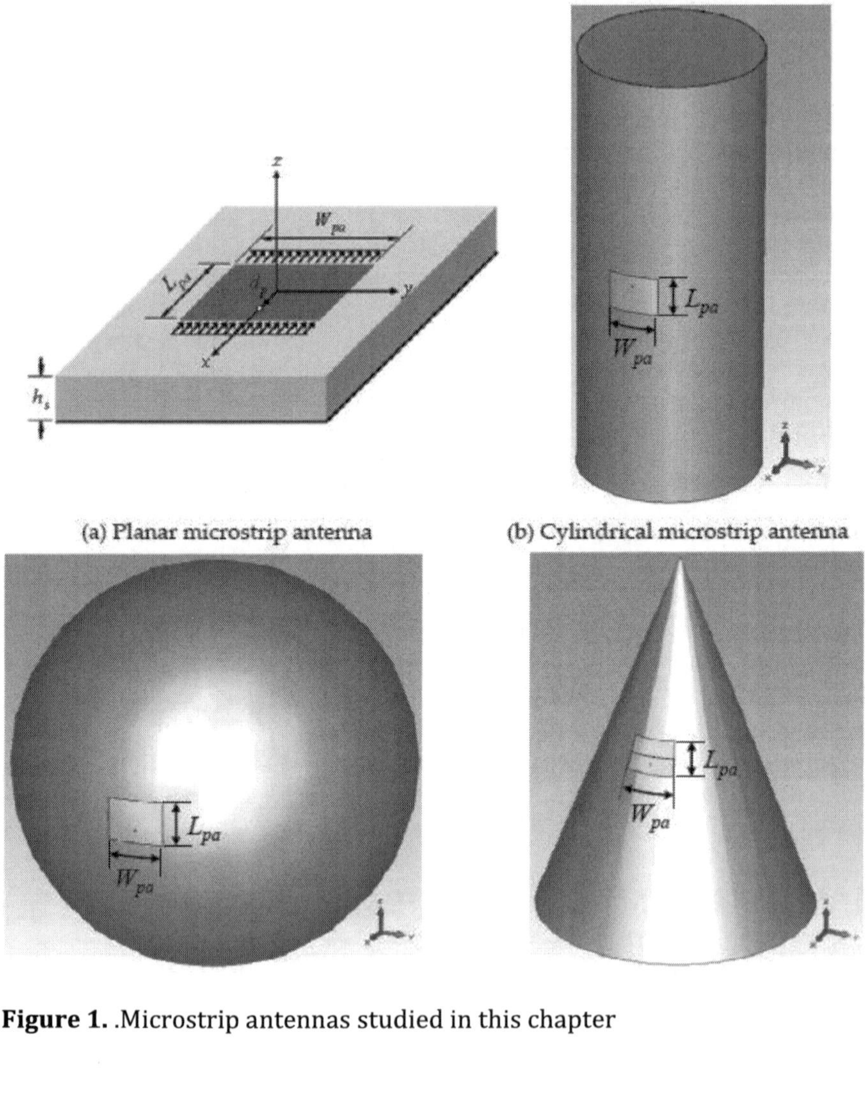

Figure 1. .Microstrip antennas studied in this chapter

2.1. Algorithm description

In order to describe the algorithm for the design of conformal microstrip antennas, for the sake of simplicity, let us first consider a probe-fed planar microstrip antenna with a gular patch of length L_{pa} and width W_{pa}, mounted on a dielectric substrate of thickness h_s, relative permittivity ε_r, and loss tangent tan δ, such as the one shown in Figure 1(a). The antenna feed probe is positioned d_p apart from the patch centre. For the following analysis, it is adopted that the antenna resonant frequency f_r is controlled by the length L_{pa} and once the probe is located along the *x*-axis, it excites the TM_{10} mode, whose main fringing field is also represented in Figure 1(a). Despite this geometry being of planar type, the same model parameters are used to describe the conformal quasi-rectangular microstrip antennas illustrated in Figure 1(b), 1(c) and 1(d), and consequently the algorithm is valid as well.

It is convenient to write both the probe position d_p and patch width W_{pa} as functions of the patch length L_{pa}, to establish a standard set of control variables. Hence, the probe position is written as

$$d_p = R_p L_{pa}, 0 < R_p \leq 0.5, \tag{1}$$

and the patch width as follows

$$W_{pa} = R L_{pa}, R \geq 1. \tag{2}$$

Therefore, the standard set of control variables is composed of L_{pa}, R(patch width to patch length ratio) and R_p (probe position to patch length ratio). The variables L_{pa} and R_p will be used in the algorithm to control its convergence and the variable R will be defined by specification, based on the desired geometry (rectangular, square). Usually, W_{pa} is made 30% higher than L_{pa}, i.e., $R = 1.3$ [14].

In this work, it is considered that the resonant frequency f_r occurs when the magnitude of the antenna reflection coefficient reaches its minimum value. Under this assumption,

$$|\Gamma_a(f_r)| = \min_f |\Gamma_a(f)|, \text{ for } f \in [f_1, f_2], \tag{3}$$

in which $\Gamma_a(f)$ is the reflection coefficient determined in the antenna full-wave analysis, f_1 and f_2 are the minimum and maximum frequencies that define the simulation domain [f_1, f_2]. For electrically thin radiators it is usually enough to choose $f_1 = 0.95 f_0$ and $f_2 = 1.05 f_0$, where f_0 is the desired operating frequency, and whether the microstrip antenna is electrically thick, then $f_1 = 0.80f_0$ and $f_2 = 1.20f_0$, in order to locate f_r between f_1 and f_2 in the first algorithm iteration.

Since the antennas design will be conducted in an iterative manner, the optimization process of the model needs to be evaluated against optimization goals in order to set a stop criterion. Therefore, let the frequency error be defined as

$$e = \left| \frac{f_r}{f_0} - 1 \right| \tag{4}$$

and its maximum value specified as e_{max}. It leads to the first optimization goal, that is,

$$e \leq e_{max} \, . \tag{5}$$

The second optimization goal is expressed by means of

$$\left| \Gamma_a(f_r) \right| < \Gamma_{min} \, , \tag{6}$$

where Γ_{min} is a positive real number defined by specification. So, the maximum reflection coefficient magnitude observed at the resonant frequency needs to be lower than Γ_{min}.

Now that the main parameters of the design algorithm have been derived, let us focus on the Adaptive Transmission Line Model, depicted in Figure 2. As can be seen, this circuital model is composed of two microstrip lines, μS_1 and μS_2, whose widths are equal to W_{pa}, an ideal transmission line TL_p – with characteristic impedance $Z_p$and electrical length $\angle E_l$ (in degrees) given by

$$\angle E_l = 360 h_s \left(\frac{c_0}{f \sqrt{\varepsilon_r}} \right)^{-1} , \tag{7}$$

where c_0 is the speed of the light in free-space –, a capacitor C, and two load terminations L_s. The ideal transmission line together with the capacitor C were included in the model to account for the impedance frequency shift due to the feed probe. In order to fit the input impedance of this model to the one determined in the antenna full-wave analysis, the reflection coefficients at the terminals of the loads L_s are written as

$$\Gamma_f(f) = (a_0 + a_1 f)e^{-j(b_0 + b_1 f)}, \tag{8}$$

in which Γ_f (f) is the reflection coefficient of the equivalent slot of impedance Z_f, and a_0, a_1, b_0, b_1 as well as Z_p and C are the set of parameters that determine the frequency response of the circuital model. It is worth mentioning that this ATLM is valid only if its variables L_{pa} and W_{pa} are kept identical to the ones used in the full-wave analysis.

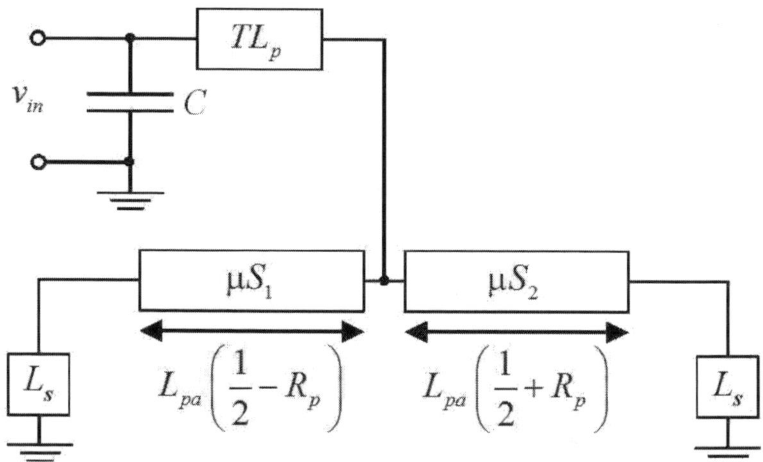

Figure 2. Adaptive transmission line model – ATLM

Once the full-wave simulation $\Gamma_a(f)$ is known, the antenna input impedance Z_a (f) can be easily evaluated. The same is valid for the circuital model analysis in which the reflection coefficient is $\Gamma_c(f)$ and input impedance is $Z_c(f)$. It is important to point out that $\Gamma_a(f)$ data can be exported from the full-wave simulator to the circuit simulator in *Touchstone* format, so $Z_a(f)$ can be utilized by the circuit simulator. The ATLM parameters set is calculated in order to have $\Gamma_c(f) = \Gamma_a(f)$ over the simulation domain $[f_1, f_2]$. The process of finding the values of this parameters set is called ATLM synthesis and it is done with aid of a Gradient optimization tool, usually available in circuit simulators such as Agilent ADS® [15], as follows.

Consider the generalized load reflection coefficient [16] that is written as

$$\Gamma_L = \frac{Z_L - Z_g^*}{Z_L + Z_g}, \tag{9}$$

in which Z_L is the load impedance and Z_g is the generator impedance, with the superscript * denoting the complex conjugate operator. Since for the ATLM the input voltage v_{in} comes from a generator, it follows that $Z_L = Z_c(f)$. By using a Gradient optimization tool with the goal $\Gamma_L = 0$ yields

$$Z_c(f) = Z_g^*, \tag{10}$$

after the optimization process.

As we want to ensure $\Gamma_c(f)=\Gamma_a(f)$, i.e., $Z_c(f)=Z_a(f)$, yields

$$Z_g = Z_a^*(f), \tag{11}$$

which is the generator impedance utilized during the ATLM synthesis. On the other hand, for the circuital simulation afterwards, $Z_g=Z_0$, where Z_0 is the characteristic impedance of the antenna feed network.

Besides, to find a meaningful solution from a physical standpoint, the following two constraints are ensured during the ATLM synthesis

$$\mathrm{Re}\{Z_f\} > 0 \text{ and } \mathrm{Im}\{Z_f\} < 0. \tag{12}$$

The complete probe-fed microstrip antenna design algorithm is depicted through the flowchart in Figure 3, which can be summarized as follows: perform a full-wave antenna simulation for a given patch length and probe position at a certain frequency range (simulation domain), which results in accurate impedance locus data; synthesize the ATLM based on the most updated full-wave simulation data available; optimize the probe position in order to match the antenna to its feed network through circuital simulation and evaluate the resonant frequency; perform patch length scaling; update the full-wave model with the new values of patch length and probe position; and repeat the whole process in an iterative manner until the goals are satisfied.

Generally, it is difficult to get the input impedance of the circuital model perfectly matched to the one obtained from full-wave simulation over the entire simulation domain $[f_1, f_2]$ (i.e., $Z_c(f) \equiv Z_a(f)$), so it is convenient to set the following goal in the Gradient optimizer,

$$\left|\Gamma_L\right| \le \begin{cases} -30\,\mathrm{dB}, f \in \left[(f_0 - \frac{f_2 - f_1}{4}), (f_0 + \frac{f_2 - f_1}{4}) \right] \\ -20\,\mathrm{dB}, f \notin \left[(f_0 - \frac{f_2 - f_1}{4}), (f_0 + \frac{f_2 - f_1}{4}) \right] \end{cases} \quad (13)$$

The previous goal contributes to reduce the number of iterations required by the Gradient optimization tool to determine the set of parameters. It was found that, in general, the required time for the synthesis of the ATLM is at most 5% of the time spent for one full-wave simulation.

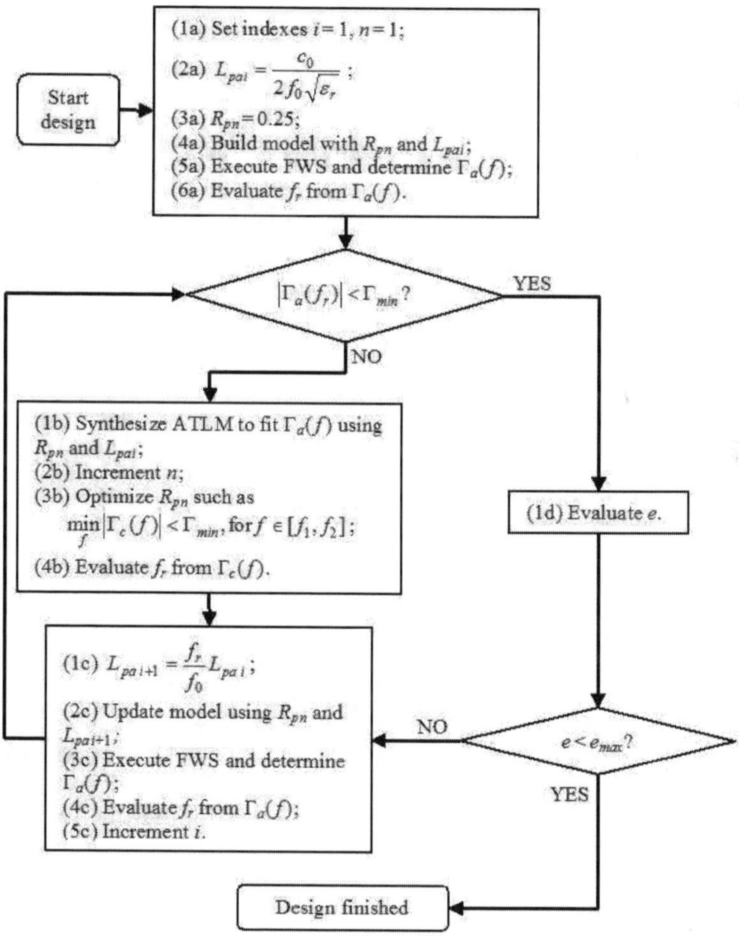

Figure 3. Probe-fed microstrip antenna design algorithm; FWS – Full-Wave Simulation

Regarding the probe position optimization, algorithm step 3b, it can be performed manually by means of a tuning process, a usual feature found in circuit simulators. Thus, R_p is tuned in order to minimize the magnitude of the input reflection coefficient of the circuital model. If desired, the optimization process can be performed employing an optimization tool, e.g., Gradient, Random, also available in circuit simulators. Usually, each circuital analysis takes no longer than 1 second using a simulator such ADS®. But, if one desires to create its own code for the ATLM circuital analysis and probe position optimization, a simple rithm can be implemented to seek the $R_p$that minimizes $|\Gamma_c(f)|$, and the computational time will be greatly reduced as well.

2.2. Applications

To illustrate the use of the technique proposed before, let us first consider the design of a cylindrical microstrip antenna (Figure 1(b)) with a quasi-rectangular metallic patch mounted on a cylindrical dielectric substrate with a thickness $h_s = 0.762$ mm, relative permittivity $\varepsilon_r = 2.5$ and loss tangent tan $\delta = 0.0022$, which covers a copper cylinder (ground layer) with a 60.0-mm radius and 300.0-mm height. The patch centre is equidistant from the top and bottom of the copper cylinder. This radiator was designed to operate at $f_0 = 3.5$ GHz and the algorithm parameters were chosen as $e_{max} = 0.1\times10^{-2}$, $\Gamma_{min} = 3.16\times10^{-2}$ (return loss of 30 dB), and $W_{pa} = 1.3\ L_{pa}$. Once it is an electrically thin antenna, the simulation domain was given by $f_1 = 0.95 f_0$ and $f_2 = 1.05 f_0$.

Following the algorithm (Figure 3), a model was built (step 4a) in the CST® software with $L_{pa1} = 27.11$ mm and $R_{p1} = 0.25$, and a first full-wave simulation was performed (step 5a). From the analysis of the obtained reflection coefficient $\Gamma_a(f)$, the determined resonant frequency was $f_r = 3.384$ GHz (step 6a) and the reflection coefficient magnitude was -17 dB, thus higher than the desired maximum of -30 dB (Figure 4(b)).

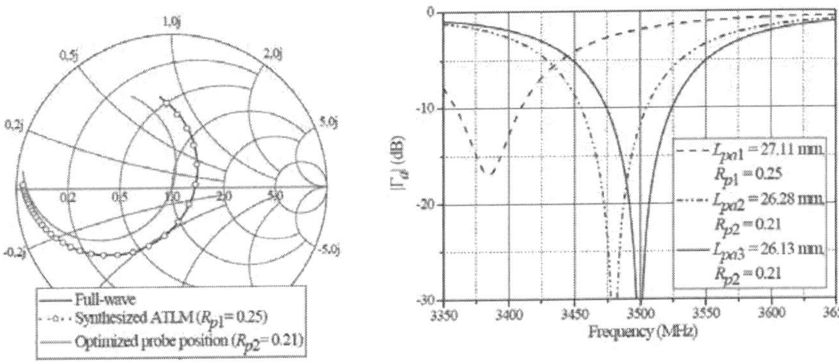

(a) (b)

Figure 4. Iterations of the algorithm for the probe-fed cylindrical microstrip antenna design: (a) impedance loci of the full-wave and circuital simulations, (b) reflection coefficient magnitude for the full-wave simulations

Hence, at the first decision point of the algorithm, the reflection coefficient magnitude at resonance is not lower than Γ_{min}, so one must go to the step 1b. Then ATLM was synthesized for L_{pa1} = 27.11 mm and R_{p1} = 0.25 and its parameters set was derived with the aid of the Gradient optimization tool of ADS®. After 55 iterations of the Gradient tool, the following parameters set was found: Z_p = 94 Ω, C = 0.87 pF, a_0 = -0.58, a_1 = 3.83×10^{-10} s, b_0 = -6.54, and b_1 = 2.21×10^{-9} s. The full-wave impedance locus and the one obtained from circuital simulation of the synthesized ATLM are shown in Figure 4(a), and it can be seen that the locus determined though circuital simulation fits very well the full-wave one.

With the circuital model available, the probe position was optimized through manual tuning of the variable R_p, and since for step 3b it is desired that the reflection coefficient magnitude at the resonance be below Γ_{min}, R_p was tuned such as the ATLM impedance locus crossed the Smith Chart centre (Figure 4(a)), leading to R_{p2} = 0.21. The resonant frequency obtained from the circuital simulation with this probe position (step 4b) was f_r = 3.392 GHz. Following the algorithm, the next step was the scaling of patch length (step 1c) leading to L_{pa2} = 26.28 mm. After updating the full-wave model with these parameters, a full-wave simulation was executed (step 3c) resulting f_r = 3.480 GHz with a reflection coefficient magnitude of -54 dB (Figure 4(b)). Since $|\Gamma_a| < \Gamma_{min}$, the next step was step 1d where it was found that e = 0.57×10^{-2}, higher than e_{max}, thus the algorithm went to step 1c, where a second patch length scaling was done leading to L_{pa3} = 26.13 mm. A last full-wave simulation with $R_{p2} = 0.21$ and $L_{pa3} = 26.13$ mm was performed resulting in $e = 0.03{\times}10^{-2}$ and return loss of 54 dB at resonance, thus satisfying all specifications. This design required only three full-wave simulations in order to guarantee all specifications, what demonstrates the efficiency of the proposed design technique.

Now let us design a probe-fed spherical microstrip antenna, such as the one illustrated in Figure 1(c). A copper sphere (ground layer) of 120.0-mm radius is covered with a dielectric substrate of constant thickness $h_s = 0.762$ mm, relative permittivity $\varepsilon_r = 2.5$ and loss tangent tan $\delta = 0.0022$. A quasi-rectangular patch with length L_{pa} and width W_{pa} is printed on the surface of the dielectric substrate. The design specifications were the same used previously and the steps of the algorithm followed a path similar to the one in the design of the cylindrical radiator. Once again, the algorithm took only three full-wave simulations to perform the design, as observed in Figure 5(a). The ATLM parameter set found was $Z_p = 91\ \Omega$, $C = 0.63$ pF, $a_0 = 6.69{\times}10^{-3}$, $a_1 = 2.32{\times}10^{-10}$ s, $b_0 = -4.10$, $b_1 = 1.54{\times}10^{-9}$ s, and the resulting patch parameters were $R_{p2} = 0.20$ and $L_{pa3} = 26.06$ mm, which led to a final frequency error $e = 0.03{\times}10^{-2}$ and 35-dB return loss at resonance.

As a last example, let us consider the design of a conical microstrip antenna with a quasi-rectangular metallic patch, as shown in Figure 1(d). It is composed of a conical dielectric substrate of constant thickness $h_s = 0.762$ mm that covers a 280.0-mm-high cone made of copper (ground layer) with a 40.0° aperture. The dielectric substrate has the same electromagnetic characteristics as the ones employed in the previous examples and the patch centre is located at the midpoint of its generatrix. This radiator was designed to operate at $f_0 = 3.5$ GHz and the algorithm parameters were chosen as $e_{max} = 0.1{\times}10^{-2}$, $\Gamma_{min} = 3.16{\times}10^{-2}$ (return loss of 30 dB), and $W_{pa} = 1.3\ L_{pa}$. By applying the developed algorithm, the ATLM parameters set found was $Z_p = 104\ \Omega$, $C = 0.33$ pF, $a_0 = -0.26$, $a_1 = 3.01{\times}10^{-10}$ s, $b_0 = -4.01$, $b_1 = 1.53{\times}10^{-9}$ s, and the determined patch parameters were $R_{p2} = 0.23$ and $L_{pa3} = 26.18$ mm, which yielded a final frequency error $e = 0.01{\times}10^{-2}$ and 34-dB return loss at resonance, once again supporting the proposed design technique. Figure 5(b) presents the reflection coefficient magnitudes of the three full-wave simulations required to accomplish the conical microstrip antenna design.

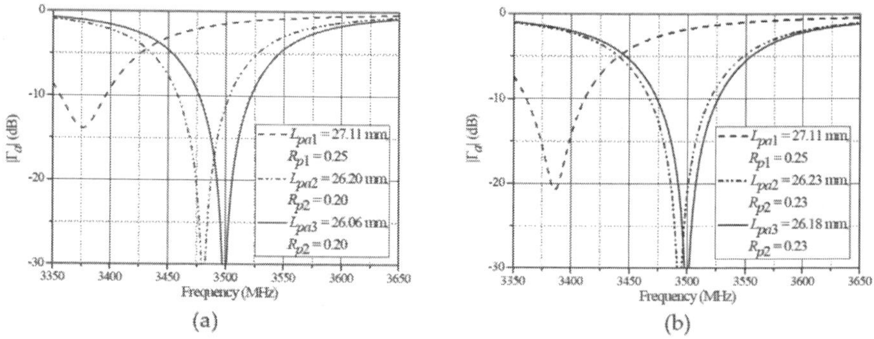

Figure 5. Reflection coefficient magnitudes for each full-wave simulation required for the designs: (a) probe-fed spherical microstrip antenna, (b) probe-fed conical microstrip antenna

3. RADIATION PATTERN SYNTHESIS OF CONFORMAL MICROSTRIP ARRAYS

The previous section addressed a computationally efficient algorithm for assisting the design of probe-fed conformal microstrip antennas with quasi-rectangular patches. In order to demonstrate its applicability, three conformal microstrip antennas were synthesized: a cylindrical, a spherical and a conical one. According to what was observed, the algorithm converges very fast, what expedites the antennas' design time.

Another concern in the design of conformal radiators is how to determine the current excitations of a conformal microstrip array to synthesize a desired radiation pattern, in which both the main beam position and the sidelobes levels can be controlled. This section is dedicated to the presentation of a technique employed for the design of conformal microstrip arrays. It is based on the iterative solution of linearly constrained least squares problems [12], so it has closed-form solutions and exhibits fast convergence, and, more important, it takes the radiation pattern of each array element into account in its code, what improves its accuracy. These radiation patterns are determined from the output data obtained through the conformal microstrip array analysis in a full-wave electromagnetic simulator, such as CST® and HFSS®. Once those data are available, polynomial interpolation is utilized to write simple closed-form expressions that represent adequately the far electric field radiated by each array element, which makes the technique numerically efficient.

The developed design technique was implemented in the Mathematica® platform giving rise to a computer program – called CMAD (Conformal Microstrip Array Design) – capable of performing the design of conformal microstrip arrays. The Mathematica® package, an integrated scientific computer software, was chosen mainly due to its vast collection of built-in functions that permit implementing the respective algorithm in a short number of lines, in addition to its many graphical resources. At the end of the section, to illustrate the CMAD ability to synthesize the radiation pattern of conformal microstrip arrays, the synthesis of the radiation pattern of three conformal microstrip array topologies is considered. First, a microstrip antenna array conformed onto a cylindrical surface is analysed. Afterwards, a spherical microstrip array is studied. Finally, the synthesis of the radiation pattern of a conical microstrip array is presented.

3.1. Algorithm description

The far electric field radiated by a conformal microstrip array composed of N elements and embedded in free space, assuming time-harmonic variations of the form $e^{j\omega t}$, can be written as

$$E = \mathbb{C}\frac{e^{-jk_0 r}}{r} I^t \cdot v(\theta, \phi), \tag{14}$$

where the constant C

is dependent on both the free-space electromagnetic characteristics, μ_0 and ε_0 , and the angular frequency ω, $k_0 = \omega \{\mu_0 \varepsilon_0\}^{1/2}$ is the free-space propagation constant,

$$I^t = \begin{bmatrix} I_1 & \cdots & I_N \end{bmatrix}, \tag{15}$$

with I_n ,$1 \leq n \leq N$, representing the current excitation of the n-th array element and the superscript t indicates the transpose operator,

$$\mathbf{v}(\theta,\phi) = \begin{bmatrix} g_1(\theta,\phi) \\ \vdots \\ g_N(\theta,\phi) \end{bmatrix}, \tag{16}$$

in which $g_n(\theta,\phi)$, $1 \leq n \leq N$, denotes the complex pattern of the n-th array element evaluated in the global coordinate system. Boldface letters represent vectors throughout this chapter.

Based on (14), the radiation pattern of a conformal microstrip array can be promptly calculated using the relation

$$| I^t \cdot \boldsymbol{v}(\theta,\phi) |^2 = w^\dagger \cdot [\boldsymbol{v}(\theta,\phi) \cdot \boldsymbol{v}^\dagger(\theta,\phi)] \cdot w, \tag{17}$$

where the complex weight w is equal to I^*, the superscript * represents the complex conjugate operator and † indicates the Hermitian transpose (complex conjugate transpose operator). Therefore, the radiation pattern evaluation requires the knowledge of both complex weight w and vector $\boldsymbol{v}$ (θ ,ϕ).

Once the array elements are chosen and their positions are predefined, to determine the vector v (θ, ϕ) tor $\boldsymbol{v}$ (θ , ϕ) it is necessary to calculate the complex patterns $\boldsymbol{g}_n$ (θ , ϕ), $1 \leq n \leq N$, of the array elements. For conformal microstrip arrays there are some well-known techniques to accomplish this [1], for example, the commonly used electric surface current method [17-19]. However, when this technique is employed to analyse cylindrical or conical microstrip arrays, for instance, it cannot deal with the truncation of the ground layer and the diffraction at the edges of the conducting surfaces that affect the radiation pattern. Moreover, the expressions derived from this method for calculating the radiated far electric field frequently involve Bessel and Legendre functions. Nevertheless, as extensively reported in the literature [20], the evaluation of these functions is not fast and requires good numerical routines. Hence, to overcome these drawbacks and to get more accurate results, in this chapter, the complex patterns $\boldsymbol{g}_n$ (θ , ϕ) are determined from the data obtained through the conformal microstrip array analysis in the CST® package. It is important to point out that other commercial 3D electromagnetic simulators, such as HFSS®, can also be used to assist the evaluation of the complex patterns $\boldsymbol{g}_n$ (θ , ϕ), since they are able to take into account truncation of the ground layer and diffraction at the edges of the conducting surfaces.

From the array full-wave simulation data, polynomial interpolation is applied to generate simple closed-form expressions that represent adequately the far electric field (amplitude and phase) radiated by each array element. In this work, the degree of the interpolation polynomials is established from the analysis of the *RMSE* (root-mean-square error), which provides a measure of similarity between the interpolated data and the ones given by CST®. For the following examples the interpolation polynomials' degrees are defined aiming at a *RMSE* less than 0.02.

Considering the previous scenario, to synthesize a radiation pattern in a given plane, it just requires the determination of the current excitations I_n present in the complex weight w. Figure 6 illustrates a typical specification of a radiation pattern containing information about the main beam direction α, the intervals intervals $[\theta_a, \theta_b]$ and $[\theta_c, \theta_d]$ where the sidelobes are located as well as the maximum level R that can be assumed for them.

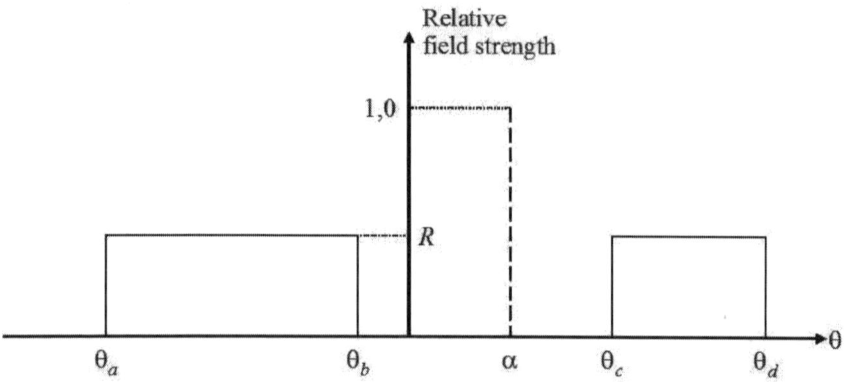

Figure 6. Typical specification of a radiation pattern in a given plane

Based on (17) and following [12], a constrained least squares problem is established in order to locate the main beam at the α direction,

$$\min_{w} w^{\dagger} \cdot A \cdot w \tag{18}$$

subject to the constraints

$$\boldsymbol{v}_s^{\dagger} \cdot w = 1, \tag{19}$$

$$\mathrm{Re}\{\boldsymbol{v}_d^{\dagger} \cdot w\} = 0, \tag{20}$$

in which $v_s = v(\alpha, \phi')$, $v_d = \partial v(\theta, \phi)/\partial\theta \,|_{(\theta\,,\,\phi) = (\alpha\,,\,\phi')}$, ϕ' is the ϕ coordinate of the plane where the pattern is being synthesized, and

$$A = \frac{1}{2}\sum_{\ell=1}^{L} \boldsymbol{v}(\theta_\ell, \phi') \cdot \boldsymbol{v}^\dagger(\theta_\ell, \phi'), \tag{21}$$

with the angles θ_ℓ, $\ell = 1, 2, \ldots, L$, uniformly sampled in the sidelobes intervals $[\theta_a, \theta_b]$ and $[\theta_c, \theta_d]$. In the next examples the adopted step size between consecutive θ_ℓ is equal to 0.1° (for each of the sidelobes intervals).

In order to find a closed-form solution to the problem defined by (18) to (20), we determine its real counterpart [21], that is,

$$\min_{\tilde{w}} \tilde{w}^t \cdot \tilde{A} \cdot \tilde{w} \tag{22}$$

subject to the following linear constraints

$$\tilde{C}^t \cdot \tilde{w} = \tilde{f}, \tag{23}$$

where

$$\tilde{w} = \begin{bmatrix} \mathrm{Re}\{w\} & \mathrm{Im}\{w\} \end{bmatrix}^t, \tag{24}$$

$$\tilde{A} = \begin{bmatrix} \mathrm{Re}\{A\} & -\mathrm{Im}\{A\} \\ \mathrm{Im}\{A\} & \mathrm{Re}\{A\} \end{bmatrix}, \tag{25}$$

$$\tilde{C} = \begin{bmatrix} \tilde{\boldsymbol{v}}_s & \hat{\boldsymbol{v}}_s & \tilde{\boldsymbol{v}}_d \end{bmatrix}, \tag{26}$$

$$\tilde{f} = \begin{bmatrix} 1 & 0 & 0 \end{bmatrix}^t, \tag{27}$$

with

$$\tilde{\boldsymbol{v}}_s = \begin{bmatrix} \mathrm{Re}\{\boldsymbol{v}_s\} & \mathrm{Im}\{\boldsymbol{v}_s\} \end{bmatrix}^t, \tag{28}$$

$$\hat{\boldsymbol{v}}_s = \begin{bmatrix} -\mathrm{Im}\{\boldsymbol{v}_s\} & \mathrm{Re}\{\boldsymbol{v}_s\} \end{bmatrix}^t, \tag{29}$$

$$\tilde{\boldsymbol{v}}_d = \begin{bmatrix} \mathrm{Re}\{\boldsymbol{v}_d\} & \mathrm{Im}\{\boldsymbol{v}_d\} \end{bmatrix}^t. \tag{30}$$

The closed-form solution to the problem (22) and (23) is

$$\tilde{w} = \tilde{A}^{-1} \cdot \tilde{C} \cdot \left(\tilde{C}^{t} \cdot \tilde{A}^{-1} \cdot \tilde{C}\right)^{-1} \cdot \tilde{f}, \tag{31}$$

from which the complex weight w is promptly evaluated.

After solving the problem (18)-(20) the main beam is located at the α-direction. Nevertheless, it cannot be assured that the sidelobes levels are below the threshold R. In order to get it, the complex weight w is updated by residual complex weights Δw, as follows:

$$w \leftarrow w + \Delta w. \tag{32}$$

A constrained least squares problem, similar to (18)-(20), that ensures the sidelobes levels, is set up for the purpose of calculating the residual complex weights Δw, that is,

$$\min_{\Delta w} \Delta w^{\dagger} \cdot A \cdot \Delta w \tag{33}$$

subject to the constraints

$$\boldsymbol{v}_s^{\dagger} \cdot \Delta w = 0, \tag{34}$$

$$\mathrm{Re}\{\boldsymbol{v}_d^{\dagger} \cdot \Delta w\} = 0, \tag{35}$$

$$\boldsymbol{v}_i^{\dagger} \cdot \Delta w = f_i, i = 1,2, \ldots, m, \tag{36}$$

in which $\boldsymbol{v}_i = \boldsymbol{v}(\theta_i, \phi')$, with θ_i denoting the θ coordinate of the i-th sidelobe, m is the number of sidelobes whose levels are being modified (the maximum m is equal to $N-2$), and the complex function f_i can be evaluated through

$$f_i = (R - |c_i|)\frac{c_i}{|c_i|}, \tag{37}$$

where

$$c_i = w^{\dagger} \cdot \boldsymbol{v}_i. \tag{38}$$

It is important to point out that the constraints (34) and (35) retain the main beam located at the α-direction, and the ones in (36) are responsible for conducting the sidelobes levels to the threshold R. A closed-form solution to the problem (33)-(36) is also determined from its real counterpart, analogous to the solution to the problem (18)-(20). The problem (33)-(36) is iteratively solved until the sidelobes levels reach the desired value R. Notice that at each iteration the maximum number of sidelobes whose levels are controlled is equal to $N - 2$, i.e., if the array radiation pattern has more than $N - 2$ sidelobes, we choose the $N - 2$ sidelobes with the highest levels to apply the constraints (36).

The radiation pattern synthesis technique described before was implemented in the Mathematica® platform with the aim of developing a CAD – called CMAD – capable of performing the design of conformal microstrip arrays. The inputs required to start the design procedure in the CMAD program are the Text Files (.txt extension) containing the points that describe the complex patterns of each array element – obtained from the conformal microstrip array simulation in CST® package –, the look direction α, the maximum sidelobes level R, and the starting and ending points of the intervals $[\theta_a, \theta_b]$ and $[\theta_c, \theta_d]$ where they are located. As a result, the CMAD returns the current excitations and the synthesized pattern. It is worth mentioning that the use of interpolation polynomials to describe the complex patterns expedites the evaluation of both vector $\boldsymbol{v}(\theta, \phi)$ and its derivative; consequently, the CMAD's run time is diminished. In the following three sections, examples of radiation pattern synthesis are provided to demonstrate the capability of the developed CMAD program.

3.2. Cylindrical microstrip array

To illustrate the described pattern synthesis technique, let us first consider the design of a five-element cylindrical microstrip array, such as the one shown in Figure 7(a). For this array, the cylindrical ground layer is made out of copper cylinder with a 60.0-mm radius and a 300.0-mm height. The employed dielectric substrate has a relative permittivity $\varepsilon_r = 2.5$, a loss tangent $\tan\delta = 0.0022$ and its thickness is $h_s = 0.762$ mm. The array patches are identical to the one designed in Section 2.2 to operate at 3.5 GHz. The five elements are fed by 50-ohm coaxial probes positioned 5.49 mm apart from the patches' centres and the interelement spacing was chosen to be $\lambda_0 / 2 = 42.857$ mm (λ_0 is the free-space wavelength at 3.5 GHz).

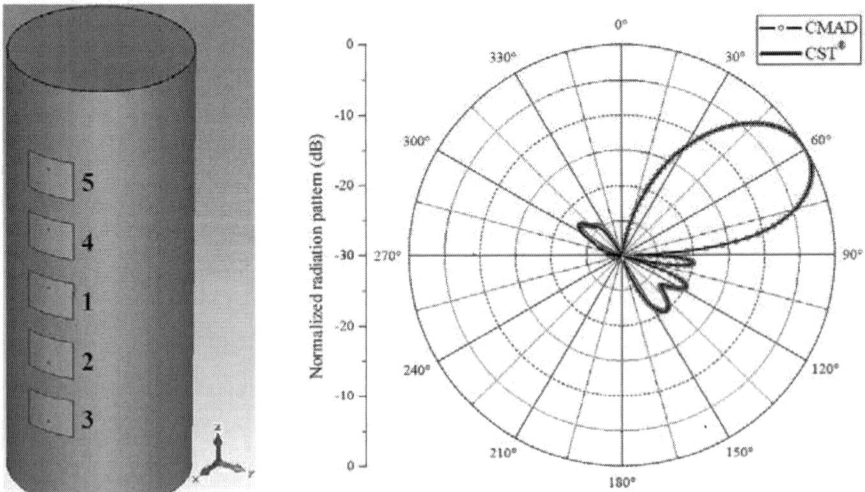

(a) (b)

Figure 7.(a)Five-element cylindrical microstrip array, (b) E_θ radiation pattern: *xz*-plane, α = 60°, *R* = -20 dB, and *f* = 3.5 GHz

It is important to point out that the elements close to the ends of the ground cylinder have significantly different radiation patterns than those close to the centre of this cylinder; however, the technique developed in this chapter can handle well this aspect, different from the common practice that assumes the elements' radiation patterns are identical [22]. To clarify this difference among the patterns, Figure 8 shows the radiation patterns of the elements number 1 and 5. In Figure 8(a) they were evaluated in CST® and in Figure 8(b) they were determined from the interpolation polynomials. As observed, there is an excellent agreement between the radiation patterns described by the interpolation polynomials and the ones provided by CST®, even in the back region, where the radiation pattern exhibits low level and oscillatory behaviour. It validates the use of polynomial interpolation functions to represent the far electric field radiated by the conformal array elements.

For this cylindrical array, let us consider that the radiation pattern in the *xz*-plane must have the main beam located at α = 60° and the maximum sidelobe level allowed is *R* = -20 dB. By using the CMAD program, we get both the array normalized current excitations, depicted in Table 1, and the synthesized radiation pattern, shown in Figure 7(b). In order to validate these results, we provide the normalized current excitations (Table 1) for the array simulation in CST®. The radiation pattern evaluated in CST® is also represented in Figure 7(b). According to what is observed, there is an excellent agreement between the radiation pattern given by the CMAD and the one calculated in CST®, thus validating the developed technique to design cylindrical microstrip arrays.

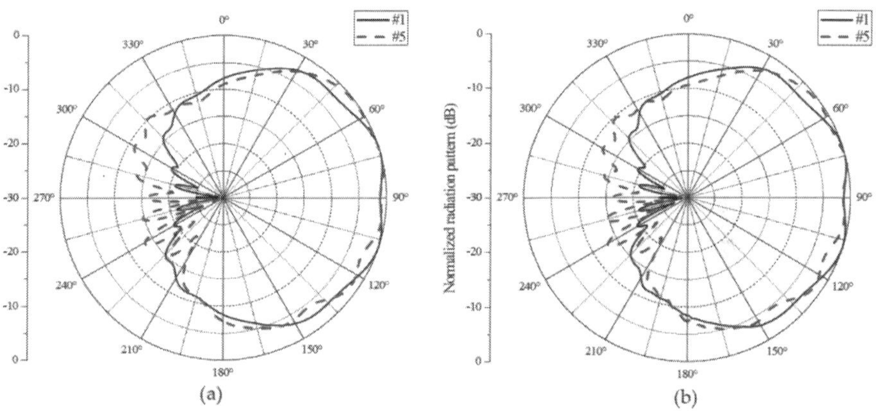

Figure 8. E_θ radiation patterns – elements number 1 and 5: *xz*-plane and f = 3.5 GHz. (a) CST® and (b) interpolation polynomials

Table 1. Cylindrical microstrip array: normalized current excitations

Element Number	Normalized Current Excitation
1	1.0∠0.0°
2	0.800∠-82.394°
3	0.360∠6.211°
4	0.781∠-90.315°
5	0.617∠172.593°

3.3. Spherical microstrip array

Another conformal microstrip array topology used to demonstrate the CMAD's ability to synthesize radiation patterns is the five-element spherical microstrip array, which operates at 3.5 GHz, illustrated in Figure 9(a). For this array, the selected ground layer is a copper sphere with a radius of 120.0 mm. A typical microwave substrate (ε_r = 2.5, tan δ = 0.0022 and h_s = 0.762 mm) covers all the ground sphere and the array patches are the same as the ones designed in Section 2.2. The angular interelement spacing in the θ-direction was chosen to be 20.334°, which corresponds to an arc length of λ_0 / 2 onto the external microwave substrate surface.

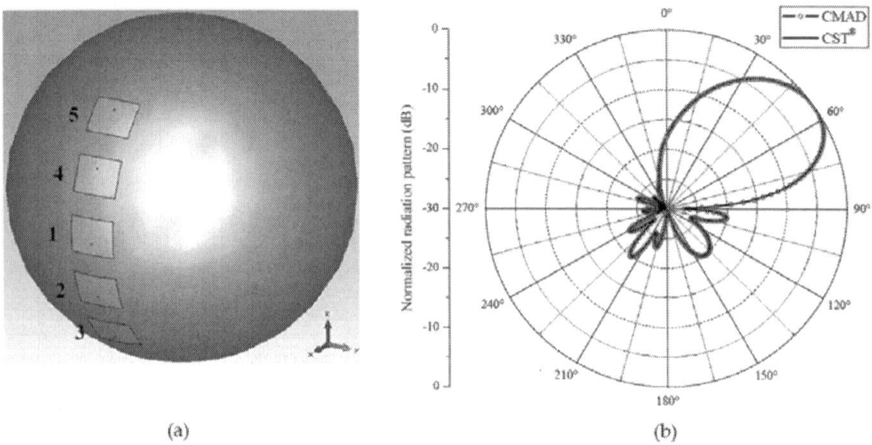

Figure 9. (a) Five-element spherical microstrip array, (b) E_θ radiation pattern: *xz*-plane, α = 55°, *R* = -20 dB, and *f* = 3.5 GHz

In this case, the synthesized radiation pattern in the *xz*-plane must have its main beam located at α = 55.0° direction and the maximum sidelobe level cannot exceed -20 dB. After entering these requirements in the CMAD program, it outputs the normalized current excitations (Table 2) and the synthesized radiation pattern (Figure 9(b)). To verify these results, the normalized current excitations were loaded into the spherical microstrip array simulation conducted in the CST® software. The radiation pattern obtained is also shown in Figure 9(b) for comparisons purposes. As seen, the radiation pattern given by the CMAD program and the one determined in CST® show a very good agreement, thus supporting the proposed radiation pattern synthesis technique. It is important to point out that the interelement spacing could be varied if the array directivity needs to be altered.

Table 2. Spherical microstrip array: normalized current excitations

Element Number	Normalized Current Excitation
1	0.680∠264.460°
2	0.252∠-6.059°
3	0.160∠156.639°
4	1.0∠0.0°
5	0.728∠-36.758°

3.4. Conical microstrip array

Finally, let us consider the radiation pattern synthesis of the four-element conical microstrip array presented in Figure 10(a). For this array, the ground layer is a 280.0-mm-high cone made of copper with a 40.0° aperture. This cone is covered with a dielectric substrate of constant thickness h_s = 0.762 mm, relative permittivity ε_r = 2.5 and loss tangent tan δ = 0.0022. The array elements are identical to the one designed in Section 2.2, so they have a length of 26.18 mm in the generatrix direction, an average width of 34.03 mm in the ϕ-direction, and the 50-ohm coaxial probes are located 6.02 mm apart from the patches' centres toward the ground cone basis. The interelement spacing in the generatrix direction is of 42.857 mm (= λ_0 / 2) as well as the centre of the element #1 is 110.0 mm apart from the cone apex in this same direction.

The radiation pattern specifications for this synthesis are: main beam direction α = 70° and maximum sidelobe level R = -20 dB, both in the *xz*-plane. By using the CMAD program, we derive the normalized current excitations, shown in Table 3, and the synthesized radiation pattern in the frequency 3.5 GHz, illustrated in Figure 10(b). Also in Figure 10(b) the array radiation pattern calculated in CST®, considering the normalized current excitations of Table 3, is presented. As observed, the radiation pattern obtained with CMAD matches the one determined in CST®, once again supporting the proposed design approach.

4. ACTIVE FEED CIRCUIT DESIGN

As can be seen, the radiation pattern synthesis technique presented in the previous section is suitable for applications that require electronic radiation pattern control, for example. However, it only provides the array current excitations, i.e., to complete the array design it is still necessary to synthesize its feed network. A simple active circuit topology dedicated to feed those arrays can be composed of branches having a variable gain amplifier cascaded to a phase shifter, both controlled by a microcontroller, and a 1 : N power divider, as depicted in Figure 11. The phase shifters play a role in controlling the phases of the current excitations, as well as the variable gain amplifiers that are responsible for settling their amplitudes. In this section, expressions for calculating the phase shifts ϕ_n and the gains G_n, in terms of the array current excitations and their electrical characteristics, including the self and mutual impedances, are derived. It is worth mentioning that the evaluated expressions take into account the mismatches between the array elements' driving impedances and the characteristic impedance Z_0 of the lines, what improves their accuracy.

At the end of this section, to illustrate the synthesis of the proposed active feed network (Figure 11), the design of the active beamformers of the three conformal microstrip arrays (cylindrical, spherical and conical) that appear along the chapter is described. Furthermore, to validate the phase shifts and gains calculated, the designed feed networks are analysed in the ADS® package.

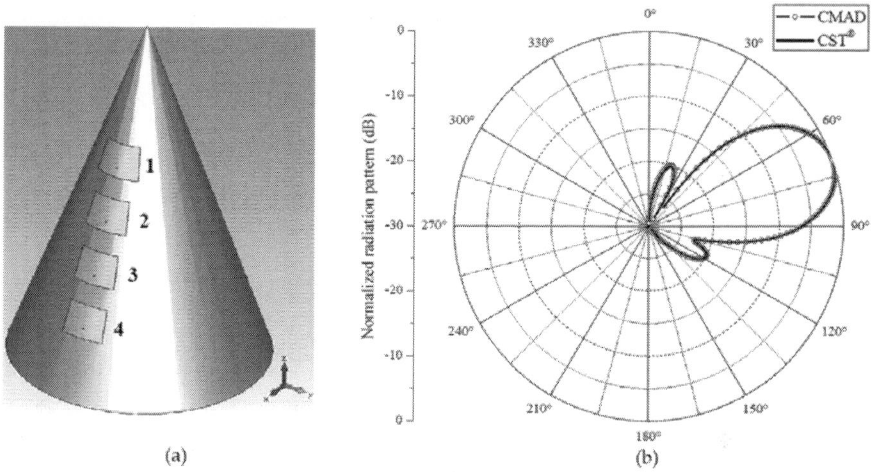

Figure 10.(a) Four-element conical microstrip array, (b) E_θ radiation pattern: *xz*-plane, α = 70°, *R* = -20 dB, and *f* = 3.5 GHz

Table 3. Conical microstrip array: normalized current excitations

Element Number	Normalized Current Excitation
1	0.574∠-7.835°
2	0.875∠0.149°
3	1.0∠0,0°
4	0.625∠-6.561°

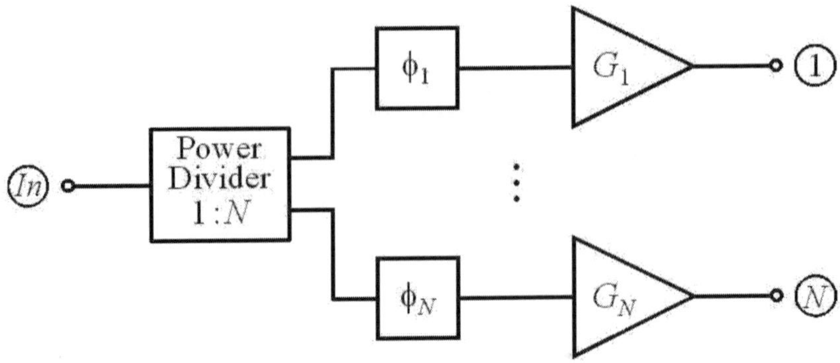

Figure 11. Active feed network

4.1. Design equations

For the analysis conducted here the phase shifters are considered perfectly matched to the input and output lines and produce zero attenuation. Based on these assumptions the scattering matrix $(S_{pf})_n$ of the n-th phase shifter, $1 \leq n \leq N$, assumes the form

$$(S_{pf})_n = \begin{pmatrix} 0 & e^{j\phi_n} \\ e^{j\phi_n} & 0 \end{pmatrix}, \tag{39}$$

with ϕ_n $(0 \leq \phi_n < 2\pi)$ representing the phase shift produced by the n-th phase shifter. Notice that the matching requirement can be met to within a reasonable degree of approximation for commercial IC (Integrated Circuit) phase shifters, however those devices frequently exhibit moderate insertion loss. So, to take the insertion loss into account in our analysis model, the gains G_n are either decreased or increased (to compensate the insertion losses).

The variable gain amplifiers are also considered perfectly matched to the input and output lines and they are unilateral devices, i.e., $s_{12n} = 0$. Hence, the scattering matrix $(S_a)_n$ of the n-th variable gain amplifier is given by

$$(S_a)_n = \begin{pmatrix} 0 & 0 \\ s_{21n} & 0 \end{pmatrix}, \tag{40}$$

in which s_{21n} denotes the gain (linear magnitude) of the n-th variable gain amplifier. It is worth mentioning that lots of commercial IC variable gain amplifiers have input and output return loss better than 10 dB and exhibit high directivity, therefore, the preceding assumptions are reasonable. More precise results using the scattering parameters of commercial variable gain amplifiers and phase shifters are presented in [23].

Let us examine the operation of the n-th circuit branch. The input power P_n at the terminals of the n-th array element can be calculated by

$$P_n = \frac{1}{2}\mathrm{Re}\{Zin_n\}\,|I_n|^2, \tag{41}$$

where Zin_n is the driving impedance at the terminals of the n-th array element and can be evaluated using

$$Zin_n = \sum_{\kappa=1}^{N} Z_{n\kappa}\frac{I_\kappa}{I_n} \tag{42}$$

in which $Z_{n\kappa}$ is the n-th array element self-impedance, if $n = \kappa$, and the mutual impedance between the n-th and κ-th array elements, if $n \neq \kappa$. In this chapter the self and mutual impedances will be determined from the array simulation data. However, those impedances could also be obtained from the measurements conducted in the array prototype, what certainly would lead to a more accurate feed network design.

Alternatively, the input power at the terminals of the n-th array element can be expressed in terms of the incident power P_{0n} and the reflection coefficient Γin_n at the terminals as

$$P_n = P_{0n}(1 - |\Gamma in_n|^2), \tag{43}$$

with

$$\Gamma in_n = \frac{Zin_n - Z_0}{Zin_n + Z_0}. \tag{44}$$

Combining (41) and (43) results in an expression to evaluate the incident power at the terminals of the n-th array element

$$P_{0n} = \frac{\mathrm{Re}\{Zin_n\}\,|I_n|^2}{2(1 - |\Gamma in_n|^2)}, \tag{45}$$

which is equal to the n-th variable gain output power, disregarding the losses in the lines.

Based on (45), an equation to determine the gain of the n-th variable gain amplifier is derived:

$$G_n = \frac{P_{0n}}{P_{0m}} = \frac{\mathrm{Re}\{Zin_n\}}{\mathrm{Re}\{Zin_m\}} \frac{1-|\Gamma in_m|^2}{1-|\Gamma in_n|^2} \frac{|I_n|^2}{|I_m|^2}. \tag{46}$$

Notice that to evaluate (46) it is necessary to choose one of the circuit branches as a reference, i.e., the gain of the m-th variable gain amplifier is set equal to 1.0.

It is important to highlight that this formulation has relevant importance for arrays whose mutual coupling among elements is strong [23], since it takes this effect into account. For arrays whose mutual coupling among elements is weak and the array elements self-impedances are close to Z_0, (46) is approximated by

$$G_n \cong \frac{|I_n|^2}{|I_m|^2}. \tag{47}$$

Now, to determine the phase shifts ϕ_n, let us consider the current I_n at the terminals of the n-th array element, that is,

$$I_n = I_{0n} e^{j\phi_n} (1 - \Gamma in_n), \tag{48}$$

in which $I_{0n} e^{j\phi_n}$ is the incident current wave at the terminals of the n-th array element.

Once the currents I_n are provided by the algorithm described in the last section, to calculate ϕ_n the phases of the left and right sides of (48) are enforced to be equal. Then,

$$\phi_n = \delta_{nm} - \arg\{1 - \Gamma in_n\} + \arg\{1 - \Gamma in_m\}, \tag{49}$$

with

$$\delta_{nm} = \arg\{I_n\} - \arg\{I_m\}. \tag{50}$$

Also for the determination of the phase shift ϕ_n, the m-th circuit branch was taken as a reference, i.e., its phase shifter does not introduce any phase shift ($\phi_m = 0°$) in the signal.

For arrays whose mutual coupling among elements is weak and the array elements self-impedances are close to Z_0, the phase shift ϕ_n (49) reduces to

$$\phi_n \cong \delta_{nm} . \tag{51}$$

The expressions for evaluating the gains G_n (46) and phase shifts ϕ_n (49) were incorporated into the developed computer program CMAD to generate a new module devoted to design active feed networks, such as the one illustrated in Figure 11. The inputs required to start the circuit design are the array current excitations and the *Touchstone* File (.sNp extension) containing the array scattering parameters – obtained from the conformal microstrip array simulation in a full-wave electromagnetic simulator, for example. In the next section, to demonstrate the capability of this new CMAD feature, the feed networks of the three conformal microstrip arrays previously synthesized will be designed.

4.2. Examples

The normalized current excitations found in Tables 1 to 3 and the scattering parameters of the three conformal microstrip arrays synthesized in this chapter (evaluated in CST®) were provided to the CMAD. As results, it returned the gains and phase shifts of the active feed networks that implement the radiation patterns shown in Figures 7(b), 9(b) and 10(b). These values are listed in Table 4.

To verify the validity of the results found in Table 4, the designed active feed networks were analysed in the ADS® package. As an example, Figure 12 shows the simulated feed network for the conical microstrip array. In this circuit, the array is represented through a 4-port microwave network, whose scattering parameters are the same as the ones used by the CMAD, it is fed by a 30-dBm power source with a 50-ohm impedance, and there are four current probes to measure the currents at the terminals of the 4-port microwave network, which correspond to the array current excitations. Table 5 summarizes the current probes readings for the three analysed feed networks. The comparison between the currents given in Table 5 and the ones presented in Tables 1 to 3 shows that these currents are in agreement, thereby validating the design equations derived before.

Table 4. Gains and phase shifts of the designed active feed networks

Branch Number	Cylindrical Array		Spherical Array		Conical Array	
	Gain (dB)	Phase Shift (deg)	Gain (dB)	Phase Shift (deg)	Gain (dB)	Phase Shift (deg)
1	7.4	351.1	11.1	114.6	0.0	0.0
2	5.7	267.6	3.8	205.0	4.2	7.0
3	0.0	0.0	0.0	0.0	5.4	4.1
4	4.9	260.6	14.0	210.3	1.5	356.6
5	2.4	164.0	10.9	173.0	–	–

Table 5. Current probes readings (in ampere)

Branch Number	Cylindrical Array	Spherical Array	Conical Array
1	0.160∠-10.37°	0.249∠113.2°	0.106∠-9.417°
2	0.128∠-92.76°	0.092∠-157.3°	0.161∠-1.432°
3	0.058∠-4.156°	0.059∠5.375°	0.184∠-1.581°
4	0.125∠-100.7°	0.366∠-151.3°	0.115∠-8.141°
5	0.099∠162.2°	0.267∠172.0°	–

5. CONCLUSION

In summary, a computationally efficient algorithm capable of assisting the design of probe-fed conformal microstrip antennas with quasi-rectangular patches was discussed. Some examples were provided to illustrate its use and advantages. As seen, it can result in significant reductions in design time, since the required number of full-wave electromagnetic simulations, which are computationally intensive – especially for conformal radiators –, is diminished. For instance, the proposed designs could be performed with only three full-wave simulations. Also in this chapter, an accurate design technique to synthesize radiation patterns of conformal microstrip arrays was introduced. The adopted technique takes the radiation pattern of each array element into account in its code through the use of interpolation

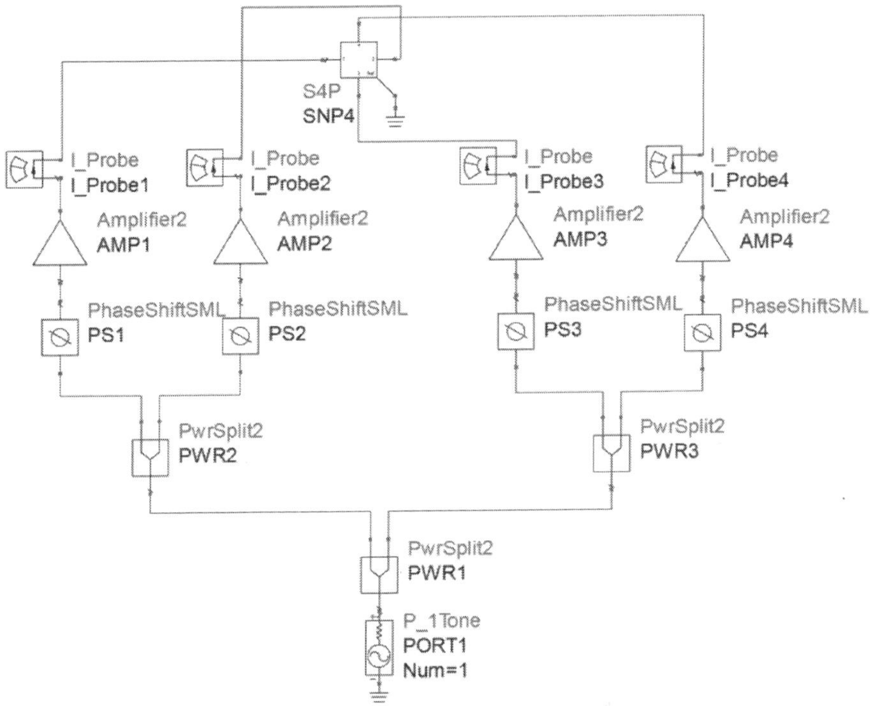

Figure 12. Simulated feed network for the conical microstrip array

polynomials, different from the common practice that assumes the elements' radiation patterns are identical. Hence, the developed technique can provide more accurate results. Besides, it is able to control the sidelobes levels, so that optimized array directivity can be achieved. This design technique was coded in the Mathematica® platform giving rise to a computer program, called CMAD, that evaluates the array current excitations responsible for synthesizing a given radiation pattern. To show the potential of the CMAD program, the design of cylindrical, spherical and conical microstrip arrays were exemplified. Finally, an active feed network suitable for applications that require electronic radiation pattern control, like tracking systems, was addressed. The expressions derived for the synthesis of this circuit take into account the mutual coupling among the array elements; therefore they are also suited for array configurations in which the mutual coupling among the elements is strong. These design equations were incorporated into the CMAD code adding to it one more project tool. In order to validate this new CMAD feature, the feed networks of the three conformal microstrip arrays described along the chapter were designed. The obtained results were validated through the feed networks' simulations in the ADS® software.

6. ACKNOWLEDGEMENTS

The authors would like to acknowledge the support given to this work, developed under the project "Adaptive Antennas and RF Modules for Wireless Broadband Networks Applied to Public Safety", with the support of the Ministry of Communications' FUNTTEL (Brazilian Fund for the Technological Development of Telecommunications), under Grant No. 01.09.0634.00 with the Financier of Studies and Projects - FINEP / MCTI.

REFERENCES

1. Wong KL. Design of Nonplanar Microstrip Antennas and Transmission Lines. New York: John Wiley & Sons, Inc.; 1999.
2. Josefsson L, Persson P. Conformal Array Antenna Theory and Design. Hoboken: John Wiley & Sons, Inc.; 2006.
3. Vasylchenko A, Schols Y, De Raedt W, Vandenbosch GAE. Quality Assessment of Computational Techniques and Software Tools for Planar-Antenna Analysis. IEEE Antennas Propagat. Magazine 2009;51(1) 23–38.
4. de Paula CB, Ferreira DB, Bianchi I. Algorithm for the Design of Linearly Polarised Microstrip Antennas. In: MOMAG 2012, 5-8 Aug. 2012, Joao Pessoa, Brazil. (in Portuguese)
5. Pues H, de Capelle AV. Accurate Transmission-Line Model for the Rectangular Microstrip Antenna. IEE Proc.-H 1984;131(6) 334–340.
6. Bhattacharyya AK, Shafai L, Gary R. Microstrip Antenna – A Generalized Transmission Line. Progress in Electromagnetics Research 1991; 4 45-84.
7. Kishk AA. Analysis of Spherical Annular Microstrip Antennas. IEEE Trans. Antennas Propagat. 1993;41(3) 338–343.

8. Richards WF, Lo YT, Harrison DD. An Improved Theory for Microstrip Antennas and Applications. IEEE Trans. Antennas Propagat. 1981;29(1) 38–46.

9. Rahmat-Samii Y, Michielssen E. Electromagnetic Optimization by Genetic Algorithms. New York: Wiley; 1999.

10. Maciel DCM. Circularly Polarised Microstrip Arrays for Beam Steering. MSc thesis. ITA – Technological Institute of Aeronautics; 2005. (in Portuguese)

11. Hansen RC. Phased Array Antennas. New York: John Wiley & Sons; 1998.

12. Tseng CY, Griffiths LJ. A Simple Algorithm to Achieve Desired Patterns for Arbitrary Arrays. IEEE Trans. on Signal Processing 1992;40(11) 2737–2746.

13. Mathematica. Wolfram Research. http://www.wolfram.com/products/mathematica/ (accessed 15 June 2012).

14. Nascimento DC, Lacava JCS. Design of Low-Cost Probe-Fed Microstrip Antennas. In: Nasimuddin N. (ed) Microstrip Antennas. Rijeka: InTech; 2011. p1-26.

15. ADS. Agilent Technologies. http://www.agilent.com/find/eesof-ads/ (accessed 15 June 2012).

16. Collin RE. Foundations for Microwave Engineering. 2nd ed. New York: McGraw-Hill; 1992.

17. Ashkenazy J, Shtrikman S, Treves D. Electric Surface Current Model for the Analysis of Microstrip Antennas on Cylindrical Bodies. IEEE Trans. Antennas Propagat. 1985;33(3) 295-300.

18. da Silva CM, Lumini F, Lacava JCS, Richards FP. Analysis of Cylindrical Arrays of Microstrip Rectangular Patches. Electron. Lett. 1991;27(9) 778-780.

19. Tam WY, Luk KM. Far Field Analysis of Spherical-Circular Microstrip Antennas by Electric Surface Current Models. IEE Proc. H Microw., Antennas Propagat. 1991;138(1) 98-102.

20. Sipus Z, Burum N, Skokic S, Kildal PS. Analysis of Spherical Arrays of Microstrip Antennas Using Moment Method in Spectral Domain. IEE Proc. Microw., Antennas and Propagat. 2006;153(6) 533-543.

21. Tseng CY. Minimum Variance Beamforming with Phase-Independent Derivative Constraints. IEEE Trans. Antennas Propagat. 1992;40(3) 285–294.

22. Allard RJ, Werner DH, Werner PL. Radiation Pattern Synthesis for Arrays of Conformal Antennas Mounted on Arbitrarily-Shaped Three-Dimensional Platforms Using Genetic Algorithms. IEEE Trans. Antennas Propagat. 2003;51(5) 1054-1062.

23. Ferreira DB, de Paula CB, Nascimento DC. Design of an Active Feed Network for Antenna Arrays. In: MOMAG 2012, 5-8 Aug. 2012, Joao Pessoa, Brazil. (in Portuguese)

2

CHAPTER

AXIAL RATIO BANDWIDTH OF A CIRCULARLY POLARIZED MICROSTRIP ANTENNA

Li Sun[1], Gang Ou[2], Yilong Lu[3] and Shusen Tan[1]

[1] Beijing Satellite Navigation Center, China

[2] College of Electronic Science and Engineering, National University of Defense Technology, China

[3] School of Electrical and Electronic Engineering, Nanyang Technological University, Singapore

1. INTRODUCTION

Microstrip antenna has been widely used due to its many advantages, such as, small volume, light weight, easy to get various polarization and easy to be integrated (Dang & Liu, 1999). Microstrip antenna can adopt many methods to obtain circular polarization (Xue and Zhong, 2002). And some technologies can achieve the miniaturization of the microstrip antenna (Xue and Zhong, 2002). Also there are some methods to enhance the impedance bandwidth of the miniaturized microstrip antenna (Liu et al., 2002) ; (Wang & Gao, 2003).

In this chapter, we focus on the axial ratio bandwidth of a circularly polarized microstrip antenna. The previous reference books discussed the axial ratio bandwidth less, always said that the axial ratio bandwidth of a circularly polarized microstrip antenna was limited, and it was less than the impedance bandwidth of a linearly polarized microstrip antenna (Lin & Nie, 2002). The group of Professor Ahmed A. Kishk has done a lot of research work on the circularly polarized microtrip antenna recently (Yang et al., 2008); (Yang et al., 2007); (Yang et al., 2006); (Chair et al., 2006); (Kishk et al., 2006). We adopt theoretical analysis and simulation by CST Microwave Studio to give out the method of improving the axial ratio bandwidth of the circularly polarized microstrip antenna.

First, we briefly introduce the basic methods which can form the circular polarization for a microstrip antenna, including the single-feed and the multiple-feed. When using multiple-feed for one patch, the sequential rotation technology (Hall et al., 1989) can be adopted. Starting from the mechanism of circular polarization obtaining from multiple-feed method, the multiple-feed can improve the axial ratio bandwidth of a microstrip antenna effectively than the single-feed microstrip antenna is demonstrated by theoretical analysis and simulation. The more feeds, the better the axial ratio bandwidth is.

Then, the detail analysis of the axial ratio bandwidth including when the amplitudes have some difference and the phase excitation of the feed point has an offset according to the designed central frequency in manufacture are described.

At last, the example of circularly polarized microstrip antenna design and test are in the section 5. Due to the volume limited in the project, we choose two feeds for the microstrip antenna.

2. CIRCULARLY POLARIZED METHOD

2.1. Simple microstrip antennas

Generally, the configuration of the simple microstrip antenna (Ung, 2007) is showed as in Fig.1. It can be simply formed by a dielectric substrate through photoetching technology or etching process. In the configuration, there are the metallic patch of certain shape on the top, the substrate layer of certain thickness and the ground plane on the bottom. The dielectric constant and the thickness of the dielectric substrate material, the shape and size of the top patch and the feeding method determine the performance of the microstrip antenna.

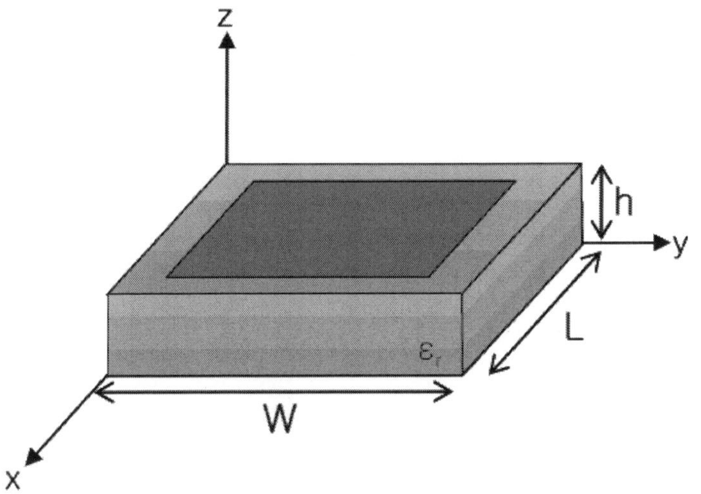

Figure 1. Configuration of the microstrip antenna

The shape of the top metallic patch can be various. Such as square, rectangle, circle, triangle, ellipse and unconventional shape, etc. The feed methods include coaxial probe feed, microstrip line feed, aperture couple feed, etc (Ung, 2007); (stutzman & Thiele, 1997). The simple microstrip antenna is usually linearly polarized. The bandwidth of the linearly polarized microstrip antenna is described by the impedance bandwidth.

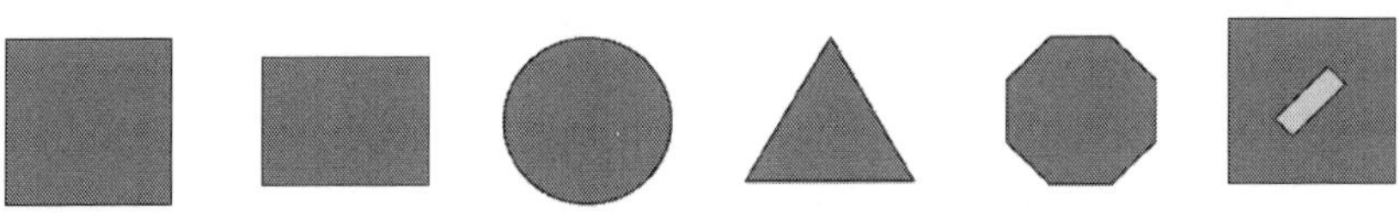

Figure 2. Patch shape

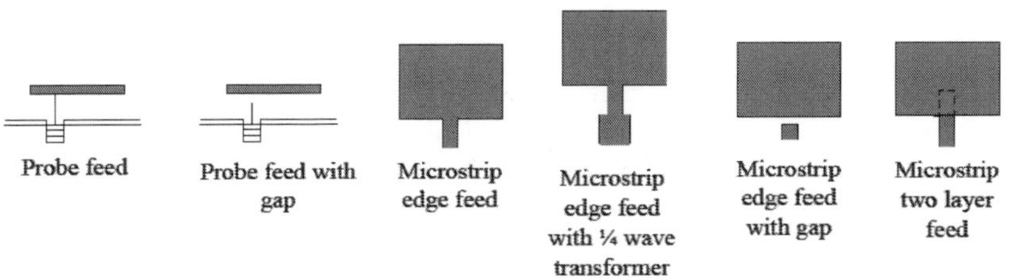

Figure 3. Feed methods

2.2. Single-feed realization method

Single-feed for the patch to form circular polarization is based on the cavity model of microstrip antenna. The two orthogonal polarized degenerate modes which can formed the circular polarization can be obtained by corner cut, quasi-square, slot, etc, and the patch shape (Lin & Nie, 2002) can be seen in Fig.4. The feed methods can adopt coaxial probe feed, aperture couple feed, etc.

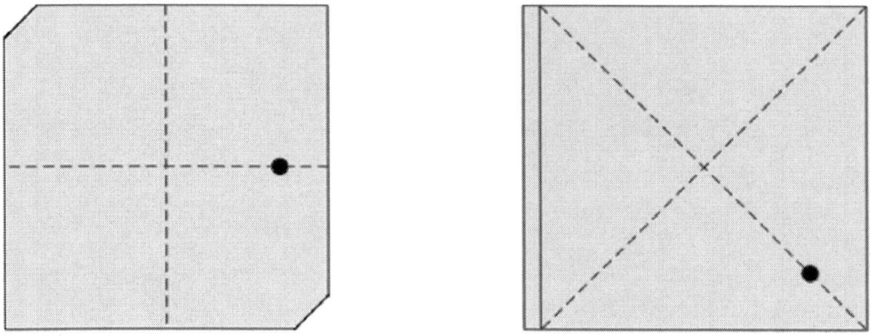

Figure 4. Patch shape of single-feed circularly polarized microstrip antenna

The axial ratio 3dB bandwidth of the circularly polarized microstrip antenna is much less than the impedance bandwidth of the linearly polarized microstrip antenna. Via application, the axial ratio 3dB bandwidth the single-feed circularly polarized microstrip antenna is limited at about 35%of the difference of the two resonant frequencies (Lin & Nie, 2002). So we must find methods to improve the axial ratio bandwidth of the circularly polarized microstrip antenna.

2.3. Multiple-feed realization method

A circularly polarized electromagnetic wave can be divided into two equal amplitudes linearly polarized components both in space and in time. Suppose that the two orthogonal polarized components are

$$\vec{E}_x = E, \quad \vec{E}_y = E e^{j\frac{\pi}{2}},$$

then we have

$$\vec{E}_y = Ee^{j\frac{\pi}{2}} = j\vec{E}_x. \tag{1}$$

Multiple-feed for one patch can adopt the sequential rotation technology. The technology of sequential rotation is successfully used in circularly polarized antenna array design (Hall et al., 1989). Multiple-feed has an appropriate phase difference between excitations, and this can improve the axial ratio bandwidth and reduce the cross-polarization. The mode exited by each feed for one patch can be regarded as the mode exited by each element in the array. So, in the case of using *M* feed points, the m_{th} feed point's phase φ_{em} can be expressed as

$$\varphi_{em} = (m-1)\frac{p\pi}{M} 1 \leq m \leq M, \tag{3}$$

where *P* is an integer.

Each feed point's physical position must have some symmetry, seen in fig.5. Through simulation, finding that fixing the first feed point position, other feed points rotate the corresponding phase differences between itself and the first feed point. The center is the disc center. In the case of *P<M*, and the last feed point does not rotate to the first feed point, it can improve axial ratio bandwidth.

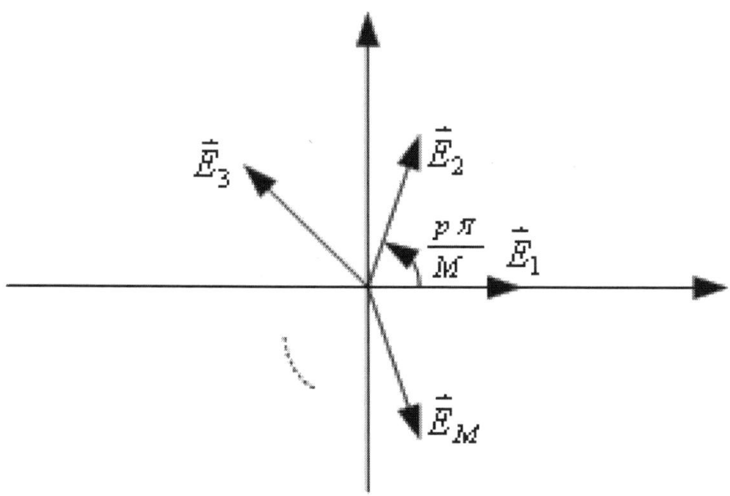

Figure 5. Feed position of multiple-feed

Suppose that

$$\vec{E}_1=E,\ \vec{E}_2=Ee^{j\frac{p\pi}{M}},\ \vec{E}_3=Ee^{j\frac{2p\pi}{M}},\ \ldots\ldots,\ \vec{E}_M=Ee^{j\frac{(M-1)p\pi}{M}}$$

so the two orthogonal components are

$$E_x=\vec{E}_1+\vec{E}_2\cos\frac{p\pi}{M}+\vec{E}_3\cos\frac{2p\pi}{M}+\cdots+\vec{E}_M\cos\frac{(M-1)p\pi}{M}$$

$$=E+Ee^{j\frac{p\pi}{M}}\cos\frac{p\pi}{M}+Ee^{j\frac{2p\pi}{M}}\cos\frac{2p\pi}{M}+\cdots+Ee^{j\frac{(M-1)p\pi}{M}}\cos\frac{(M-1)p\pi}{M}$$

$$=1+\cos^2\frac{p\pi}{M}+\cos^2\frac{2p\pi}{M}+\cdots+\cos^2\frac{(M-1)p\pi}{M}+\frac{1}{2}j\left[\sin\frac{2p\pi}{M}+\sin\frac{4p\pi}{M}+\cdots+\sin\frac{2(M-1)p\pi}{M}\right]$$

$$E_y=\vec{E}_2\sin\frac{P\pi}{M}+\vec{E}_3\sin\frac{2P\pi}{M}+\cdots+\vec{E}_M\sin\frac{(M-1)P\pi}{M}$$

$$=Ee^{j\frac{p\pi}{M}}\sin\frac{p\pi}{M}+Ee^{j\frac{2p\pi}{M}}\sin\frac{2p\pi}{M}+\cdots+Ee^{j\frac{(M-1)p\pi}{M}}\sin\frac{(M-1)p\pi}{M}$$

$$=\frac{1}{2}\left[\sin\frac{2p\pi}{M}+\sin\frac{4p\pi}{M}+\cdots+\sin\frac{2(M-1)p\pi}{M}\right]+j\left[\sin^2\frac{p\pi}{M}+\sin^2\frac{2p\pi}{M}+\cdots+\sin^2\frac{(M-1)p\pi}{M}\right]$$

According to the following formula,

$$\sum_{k=1}^{n}\sin(x+k\alpha)=\frac{\sin\frac{1}{2}n\alpha}{\sin\frac{1}{2}\alpha}\sin\left(x+\frac{1}{2}(n+1)\alpha\right),$$

we can get

$$\sin\frac{2p\pi}{M}+\sin\frac{4p\pi}{M}+\cdots+\sin\frac{2(M-1)p\pi}{M}$$
$$=\frac{\sin\frac{(M-1)p\pi}{M}}{\sin\frac{p\pi}{M}}\sin\left[\frac{1}{2}(M-1+1)\frac{2p\pi}{M}\right]=\frac{\sin\frac{(M-1)p\pi}{M}}{\sin\frac{p\pi}{M}}\sin(p\pi)=0,$$

and according to

$$\sum_{k=1}^{n}\cos(x+k\alpha)=\frac{\sin\frac{1}{2}n\alpha}{\sin\frac{1}{2}\alpha}\cos\left(x+\frac{1}{2}(n+1)\alpha\right),$$

we can get

$$\cos\frac{2p\pi}{M}+\cos\frac{4p\pi}{M}+\cdots+\cos\frac{2(M-1)p\pi}{M}$$

$$=\frac{\sin\frac{(M-1)p\pi}{M}}{\sin\frac{p\pi}{M}}\cos\left[\frac{1}{2}(M-1+1)\frac{2p\pi}{M}\right]=\frac{\sin\left(p\pi-\frac{p\pi}{M}\right)}{\sin\frac{p\pi}{M}}\cos(p\pi)=-1.$$

$$1+\cos^2\frac{p\pi}{M}+\cos^2\frac{2p\pi}{M}+\cdots+\cos^2\frac{(M-1)p\pi}{M}$$

$$=\frac{M}{2}+\frac{1}{2}+\frac{1}{2}\left(\cos\frac{2p\pi}{M}+\cos\frac{4p\pi}{M}+\cdots+\cos\frac{2(M-1)p\pi}{M}\right)=\frac{M}{2},$$

$$\sin^2\frac{p\pi}{M}+\sin^2\frac{2p\pi}{M}+\cdots+\sin^2\frac{(M-1)p\pi}{M}$$

$$=\frac{M}{2}-\frac{1}{2}-\frac{1}{2}\left(\cos\frac{2p\pi}{M}+\cos\frac{4p\pi}{M}+\cdots+\cos\frac{2(M-1)p\pi}{M}\right)=\frac{M}{2},$$

so

$$1+\cos^2\frac{p\pi}{M}+\cos^2\frac{2p\pi}{M}+\cdots+\cos^2\frac{(M-1)p\pi}{M}=\sin^2\frac{p\pi}{M}+\sin^2\frac{2p\pi}{M}+\cdots+\sin^2\frac{(M-1)p\pi}{M}=\frac{M}{2}$$

Therefore we can get

$$E_y=jE_x. \tag{3}$$

That is (1), so the multiple-feed method above has realized the circular polarization.

3. THEORETICAL ANALYSIS OF THE AXIAL RATIO BANDWIDTH

3.1. Axial ratio

We can use the polarization ellipse to describe the elliptical polarization. The instantaneous electric field orientation can figure out an ellipse in the space, seen in Fig.6.

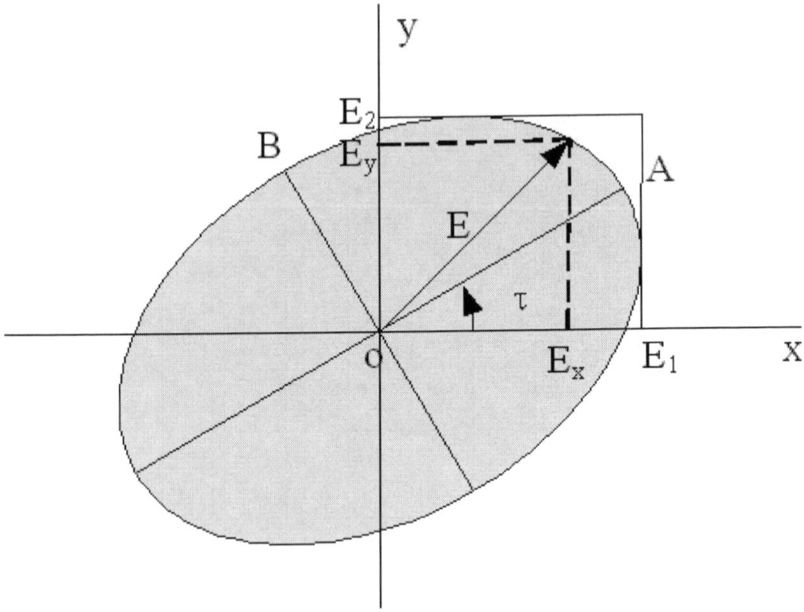

Figure 6. Polarization ellipse

The axial ratio is defined as

$$AR = {}^{OA}\!/_{OB} \; (1 \le AR \le \infty) \tag{4}$$

where *OA* is the half major axis of the polarization ellipse, and the *OB* is the half minor axis of the polarization ellipse.

The elliptical polarization of electromagnetic wave can be divided into two linearly polarized components. One's orientation is along x-axis, and the other is along y-axis. Suppose that the two linearly polarized components are $E_x = E_1 \sin(\omega t - \beta z), E_y = E_2 \sin(\omega t - \beta z + \delta)$,

where E_1 is the amplitude of the linear polarization along x-axis E_x, and E_2 is the amplitude of the linear polarization along y-axis E_y. $\delta$is the phase difference between E_x and E_y. Based on the above, we will analyze the axial ratio bandwidth of the multiple-feed microstrip antenna in the next section.

3.2. Axial ratio bandwidth of two feeds

Assume that the amplitudes excitation of each feed are equal, mutual coupling is small, and it can be neglected. Only the frequency changes the phase excitation relationship between the feed points. In the real case, usually using power splitter with separation to realize the equal amplitude excitation, and using different microstrip line length to realize the phase excitation difference. So the assumption is reasonable.

Two feeds: *M=2, P=1*. The two orthogonal electric fields are

$$E_x = E_1 \sin(\omega t - \beta z), \tag{5}$$

$$E_y = E_2 \sin(\omega t - \beta z + \delta). \tag{6}$$

At *z=0*,

$$E_x = E_1 \sin \omega t, \tag{7}$$

$$E_y = E_2(\sin \omega t \cos \delta + \cos \omega t \sin \delta), \tag{8}$$

where $\sin\omega t = E_x / E_1$, $\cos\omega t = \sqrt{1-(E_x/E_1)^2}$

Substitute (7) into (8), we can get

$$aE_x^2 - bE_xE_y + cE_y^2 = 1, \tag{9}$$

where

$$a = \frac{1}{E_1^2 \sin^2\delta}, \quad b = \frac{2\cos\delta}{E_1E_2\sin^2\delta}, \quad c = \frac{1}{E_2^2\sin^2\delta}.$$

Construct an ellipse equation

$$\frac{E_x'^2}{A^2} + \frac{E_y'^2}{B^2} = 1, \tag{10}$$

where

$$E_x' = E_x\cos\theta - E_y\sin\theta, \quad E_y' = E_x\sin\theta + E_y\cos\theta.$$

Thus (10) becomes,

$$(\frac{\cos^2\theta}{A^2}+\frac{\sin^2\theta}{B^2})E_x^2-(\frac{\sin 2\theta}{A^2}-\frac{\sin 2\theta}{B^2})E_xE_y+(\frac{\sin^2\theta}{A^2}+\frac{\cos^2\theta}{B^2})E_y^2=1. \quad (11)$$

Through (9) and (11), we can get

$$A=\sqrt{\frac{2}{a+c+\sqrt{(a-c)^2+b^2}}},B=\sqrt{\frac{2}{a+c-\sqrt{(a-c)^2+b^2}}}.$$

So

$$AR=\frac{A}{B}=\sqrt{\frac{(E_1/E_2)^2+1-\sqrt{(E_1/E_2)^4+1+2\cos 2\delta(E_1/E_2)^2}}{(E_1/E_2)^2+1+\sqrt{(E_1/E_2)^4+1+2\cos 2\delta(E_1/E_2)^2}}}. \quad (12)$$

Two feeds, when $E1/E2=1$, we can get (13) from (12).

$$AR=tg\frac{\delta}{2}. \quad (13)$$

3.3. Axial ratio bandwidth of four feeds

We analyze the axial ratio bandwidth of multiple-feed antenna, in the case of amplitude excitations are equal, and mutual coupling is neglected. Four feeds, when $M=4, P=2$. In other words, the phase excitation difference is 90°. At $z=0$, the two orthogonal electric fields are

$$E_x=E_1\sin\omega t-E_3\sin(\omega t+2\delta), \quad (14)$$

$$E_y=E_2\sin(\omega t+\delta)-E_4\sin(\omega t+3\delta), \quad (15)$$

where

$$E_y=(E_2\cos\delta-E_4\cos 3\delta)\sin\omega t+(E_2\sin\delta-E_4\sin 3\delta)\cos\omega t. \quad (16)$$

In the case of $E_1=E_2=E_3=E_4$,

$$\cos\omega t=\frac{2E_x\cos\delta-E_y}{-2E_1\sin\delta}\quad \sin\omega t=\sqrt{1-(\frac{2E_x\cos\delta-E_y}{-2E_1\sin\delta})^2}$$

Substitute into (16), we can get

$$aE_x{}^2 - bE_xE_y + cE_y{}^2 = 1, \tag{17}$$

where

$$a = \frac{1}{4E_1{}^2\sin^4\delta},$$

$$b = \frac{\cos\delta\cos2\delta}{E_1{}^2\sin^4\delta} + \frac{\cos\delta}{E_1{}^2\sin^2\delta},$$

$$c = \frac{\cos^2\delta}{4E_1{}^2\sin^4\delta} + \frac{1}{4E_1{}^2\sin^2\delta}.$$

So

$$AR = \frac{A}{B} = \sqrt{\frac{a+c-\sqrt{(a-c)^2+b^2}}{a+c+\sqrt{(a-c)^2+b^2}}} = \sqrt{\frac{1-2\cos^3\delta}{1+2\cos^3\delta}}. \tag{18}$$

That is

$$AR = \sqrt{\frac{1-2\cos^3\delta}{1+2\cos^3\delta}}. \tag{19}$$

3.4. Comparison of the two feeds and the four feeds

Next we give out the expression for phase excitation difference δ between the two feeds. The feed network substrate's relative dielectric constant is ε_r, the substrate thickness is *h*, and the width of the microstrip line is *W*. With the theory of the microstrip line, the effective dielectric constantε_{re} is (Lin & Nie, 2002)

$$\varepsilon_{re} = \frac{\varepsilon_r+1}{2} + \frac{\varepsilon_r-1}{2}(1+\frac{12h}{W})^{-\frac{1}{2}}. \tag{20}$$

The phase velocity's wavelength λ_p of the quasi-TEM wave propagated in the microstrip line is

$$\lambda_p = \frac{c}{f\sqrt{\varepsilon_{re}}}, \tag{21}$$

where c is the velocity of light in the vacuum, and f is frequency.

Assume that the microstrip line length x which providing 90° phase excitation according to the central frequency, provide δ phase excitation in fact due to the changing of the frequency, $\delta/x = 360/\lambda_p$, then

$$\delta = \frac{360xf\sqrt{\varepsilon_{re}}}{c}. \tag{22}$$

The phase excitation difference of each feed in the feed network is designed according to the central frequency. The phase excitation difference which provided by the microstrip line is changing according to the changing frequency. This will affect the circular polarization out side the central frequency.

We use the CST microwave studio to simulate the multiple-feed microstrip antenna. The simulation files are showed in Fig.7. Thorough simulation and calculation, we give out the axial ratio bandwidth comparison between two feeds and four feeds in Fig.8. Through the theoretical computation, we demonstrate that multiple-feed for one patch can effectively improve the axial ratio bandwidth. The axial ratio 3dB bandwidth of two feeds can achieve 42.6%, and four feeds can achieve 74%.

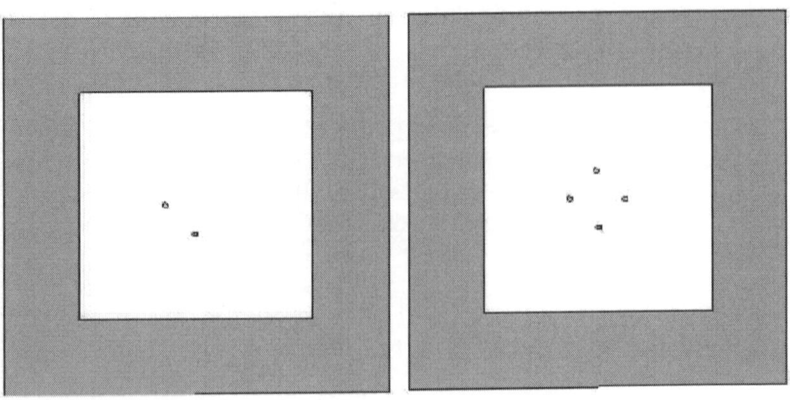

Figure 7. Simulation files of two feeds and four feeds

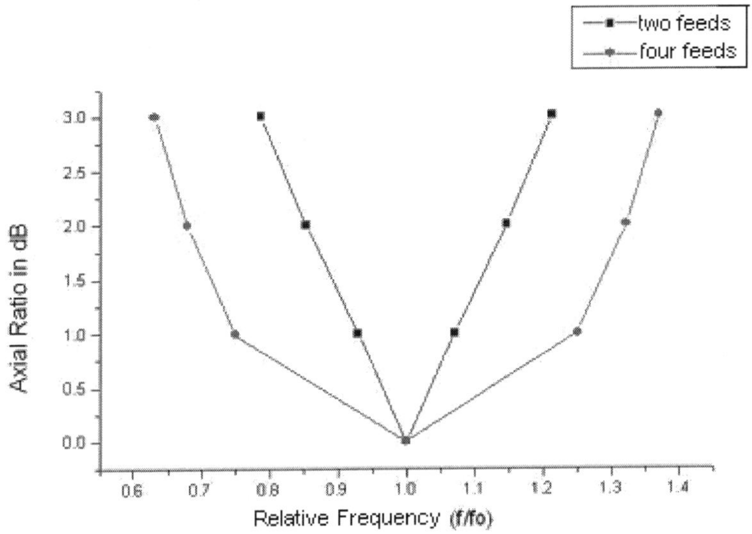

Figure 8. Axial ratio bandwidth comparison between two feeds and four feeds

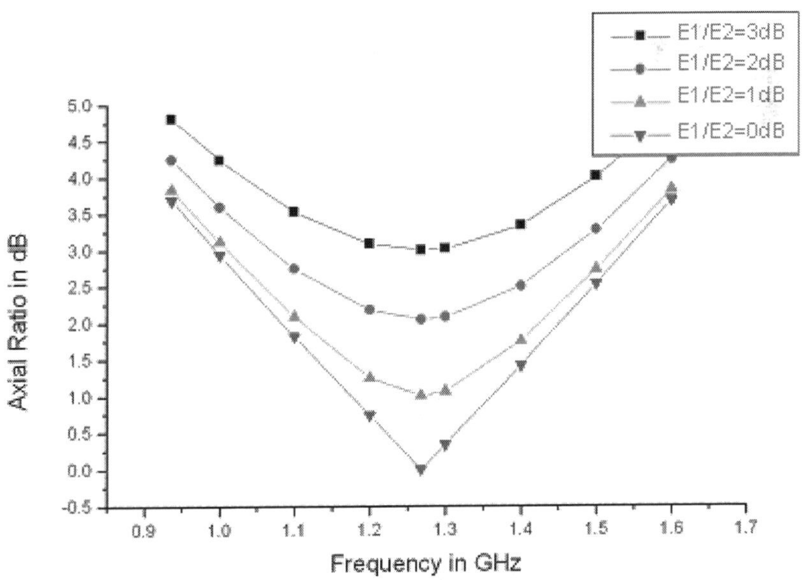

Figure 9. Axial ratio bandwidth of different amplitudes excitation of the two feeds

4. AXIAL RATIO BANDWIDTH ANALYSIS WHEN MANUFACTURE ERROR EXIST

When two feeds, assume that the amplitudes excitation are equal at every frequency. But if we substitute (22) into (12), we can get the changing of the axial ratio bandwidth according to the different ratio of E1 and E2, showing in Fig.9.

We can get the conclusion that the amplitude difference between the two feeds affects the axial ratio badly. When the amplitude ratio of the two feeds is 3dB, the axial ratio 3dB bandwidth has already disappeared.

Next we have a look at the axial ratio bandwidth changing when the phase excitation designed at the central frequency has an offset. In the feed network, change the microstrip line length x which provides 90° phase excitation to the length which provides 85.8° phase excitation. Using the same process, we can give out the changing of the axial ratio bandwidth when two feeds amplitudes are equal in Fig.10. When two feeds amplitudes ratio is 2dB in Fig.11.

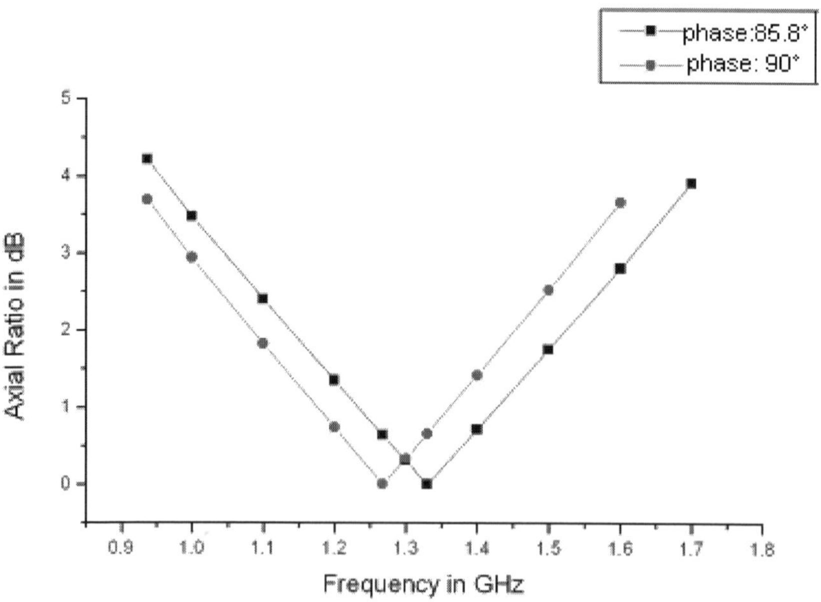

Figure 10. Axial ratio bandwidth of phase excitation has an offset at the central frequency in case of E1/E2=0dB

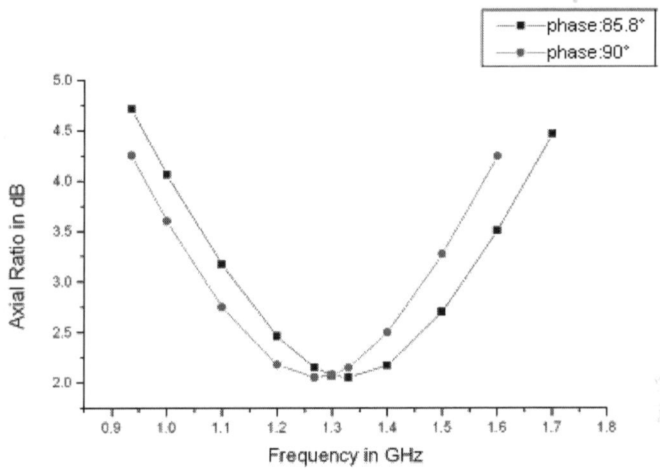

Figure 11. Axial ratio bandwidth of phase excitation has an offset at the central frequency in case of E1/E2=2dB

We can see that there is an offset on the axial ratio bandwidth when the phase excitation designed at the central frequency has an offset. From our theoretical analysis, we can get the conclusion that the multiple-feed technology can improve the axial ratio bandwidth of the microstrip antenna effectively. To get a wide band circularly polarized microstrip antenna, first, we must determine the most feed points we can use in the design according to the size limited in the project.

5. ANTENNA DESIGN EXAMPLE

5.1. Design

The more feeds, the better the axial ratio bandwidth of the circularly polarized microstrip antenna. But the feed network is more complicated and the feed network needs more space to realize.

We design a small antenna, using two feeds. Two linearly polarized components which are equal amplitude and 90°phase difference form the circular polarization. The patch shape is in Fig.12 (Hall et al., 1989), and the stubs on the patch are used to debug the resonant frequency in antenna manufacture. The feed network is in Fig.13.

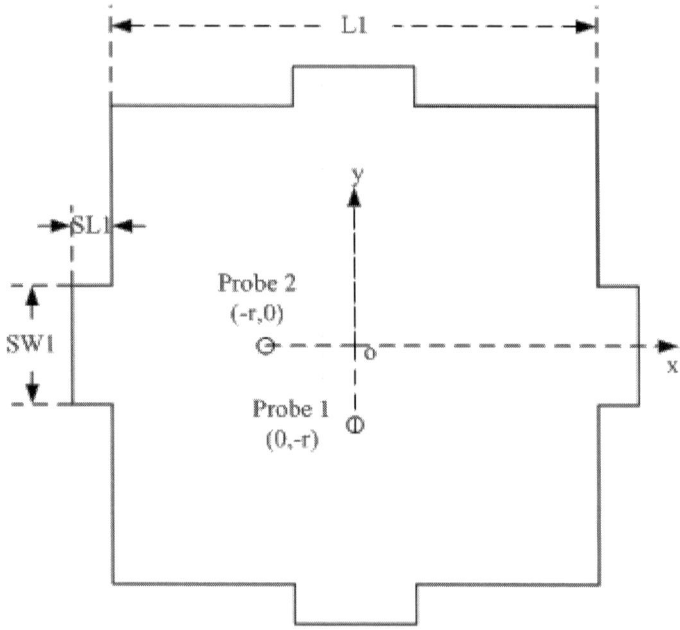

Figure 12. Patch shape

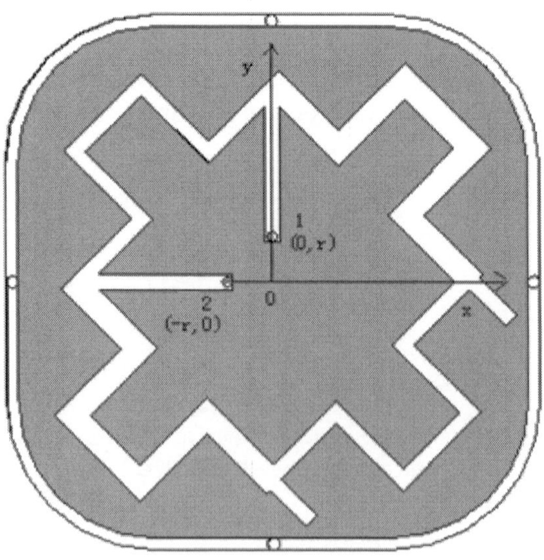

Figure 13. Feed network

5.2. Simulation analysis

Simulate the two feeds microstrip antenna we design in the above section using the CST microwave studio. We compare the difference in the axial ratio bandwidth between the single-feed and the two feeds through simulation. The configurations of the single-feed and the two feeds microstrip antenna are showed in Fig.14.

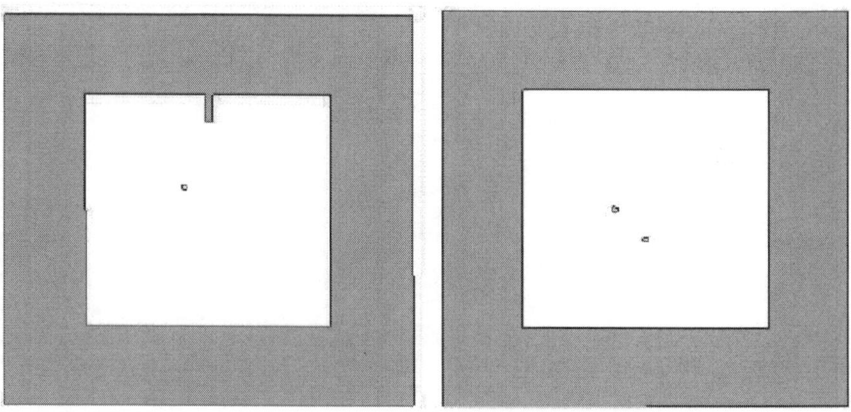

Figure 14. Simulation configuration of the single-feed and the two feeds

The simulation results of the axial ratio of the single-feed and the two feeds at zenith are showed in Fig.15. We can see that the axial ratio bandwidth of the single-feed is very limited. For the two feeds, the phase difference of the two equal amplitudes and 90° phase difference linearly polarized components according to the centre frequency change slowly and smoothly with the frequency band. This can improve the axial ratio bandwidth of a circularly polarized microstrip antenna.

5.3. Test result

The manufactured two feeds microstrip antenna is tested in the anechoic chamber. The test result of the axial ratio is showed in Fig.16.

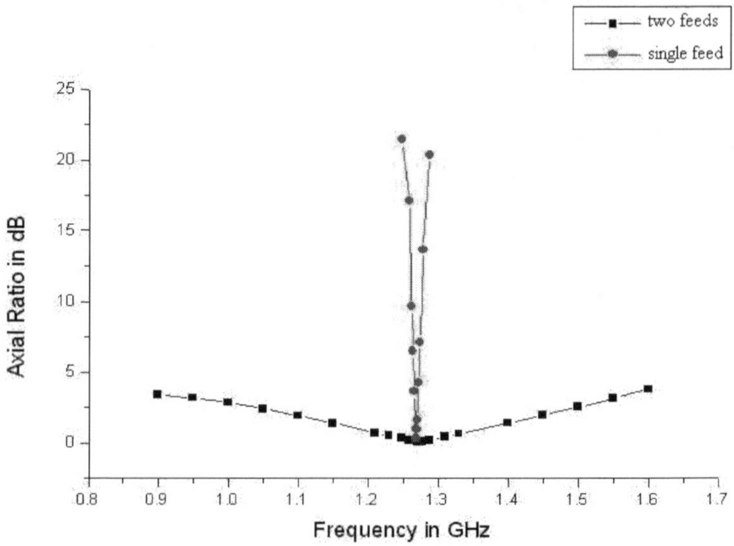

Figure 15. Axial ratio simulation results of the single-feed and the two feeds

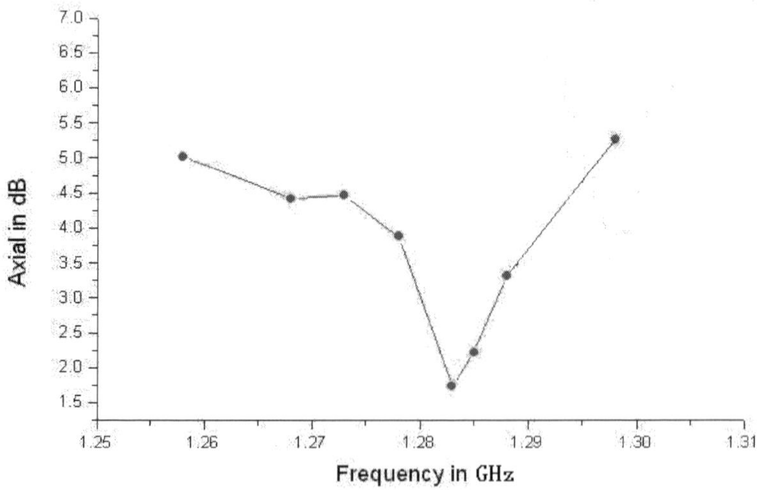

Figure 16. Axial ratio test result

In simulation, the two feeds are ideal equal amplitudes and 90° phase difference. In the manufacture, the microstrip line feed network provides the two equal amplitudes and 90° phase difference excitations. Due to the dielectric constant error of the substrate material and error of manufacture, the axial ratio bandwidth of the microstrip antenna get worse compared to the simulation result. The axial ratio 3dB bandwidth tested of the microstrip antenna is about 10MHz.

6. CONCLUSION

Microstrip antenna has been used in every field, due to its many advantages. Our main research topic in this chapter was how to improve the axial ratio bandwidth of a circularly polarized microstrip antenna. Multiple-feed method can realize the circular polarization for a microstrip antenna. Circularly polarized microstrip patch antenna designed by the multiple-feed method adopting the sequential rotation technology can improve the axial ratio bandwidth effectively. In this chapter, we demonstrate it by theoretical analysis.

Through simulation by CST Microwave Studio and theoretical computation, the axial ratio 3dB bandwidth of two feeds can achieve 42.6%, and four feeds can achieve 74%.

In engineering, choosing the most feed points according to the feed network space limited in the project can improve the axial ratio bandwidth of a circularly polarized microstrip antenna. And it is at the price of a complicated feed network compared to the few feed points design.

REFERENCES

1. H. S Dang, Y. G Liu, 1999The analysis and design of a microstrip antenna. (in Chinese). *Journal of Detection and Control*, 211March. 1999) page numbers (35-39)

2. P. S Hall, J. S Dahele, J. R James, 1989Design principles of sequentially fed, wide bandwidth, circularly polarized microstrip antennas, *Proceedings of IEE Microwaves. Antennas and Propagation*, 3813890095-0107X, 1365October, 1989, IEE

3. C. L Lin, Z. P Nie, 2002*Antenna engineering handbook*, Publishing house of electronics industry, 7-50537-495-8
4. Z. F Liu, S. W Lu, S. Z Li, 2002Improved Method for Designing Wideband Microstrip Antennas. (in Chinese). *Journal of Beijing University of Aeronautics and Astronautics*, 261February. 2000) page numbers (15-18)
5. W Stutzman, G Thiele, 1997*Antenna theory and design*, Wiley, 978-0-47102-590-0US
6. Ung Suok Kim2007Mitigation of signal biases introduced by controlled reception pattern antennas in a high integrity carrier phase differential GPS system, *dissertation*. Stanford University. March, 2007
7. C. M Wang, X. J Gao, 2003Technologies of broadband microstrip antenna. (in Chinese). *Electronic Warfare Technology*, 185September. 2003) page numbers (23-26)
8. R. F Xue, S. S Zhong, 2002Survey and progress in circular polarization technology of microstrip antennas. (in Chinese). *Chinese Journal of Radio Science*, 174August. 2002) page numbers (331-336)
9. S. S Yang, K. F Lee, A. A Kishk, K. M Luk, 2008Design and study of wideband single feed circularly polarized microstrip antennas. *Progress In Electromagnetics Research*, PIER 80, page numbers 4561
10. S. L. S Yang, A. A Kishk, K. F Lee, 2008Wideband Circularly Polarized Antenna with L-shaped Slot. *IEEE Transactions on Antennas and Propagations*, 566June. 2008) page numbers 1780-1783
11. S. L. S Yang, R Chair, A. A Kishk, K. F Lee, K. M Luk, 2007Study on Sequential Feeding Networks for Sub-Arrays of Circularly Polarized Elliptical Dielectric Resonator Antenna. *IEEE Transactions on Antennas and Propagation*, 552February. 2007) page numbers 321-333
12. S. L. S Yang, R Chair, A. A Kishk, K. F Lee, K. M Luk, 2006Single Feed Elliptical Dielectric Resonator Antennas for Circularly Polarized Applications. *Microwave and Optical Technology Letters*, 4811November. 2006) page numbers 2340-2345
13. R Chair, A. A Kishk, K. F Lee, 2006Aperture Fed Wideband Circularly Polarized Rectangular Stair Shaped Dielectric Resonator Antenna. *IEEE Transactions on Antennas and Propagations*, 544April 2006) page numbers 1350-1352

14. A. A Kishk, 2003Performance of planar four elements array of single-fed circularly polarized dielectric resonator antenna. *Microwave and Optical Technology Letters*, 385page numbers 381-384

3

CHAPTER

ON THE DESIGN OF SLOT CUT CIRCULARLY POLARIZED CIRCULAR MICROSTRIP ANTENNAS

Amit A. Deshmukh, Saleha A. Shaikh, Ami A. Desai, Kshitij A. Lele, Sudesh Agrawal

Department of Electronics and Telecommunication, Dwarkadas J. Sanghvi College of Engineering (DJSCE), Mumbai, India

ABSTRACT

Circularly polarized microstrip antenna is frequently realized by cutting the slot inside the patch and feeding it along the diagonal axis. In the reported literature, procedure to design them at any given frequency is not available. In proposed work, circularly polarized slot cut circular microstrip antenna at 900 MHz is discussed. By studying the surface current distributions at two orthogonal modes, formulations in their resonant length are proposed. The frequencies calculated using them closely agree with simulated results. Using proposed formulation, procedure to design circular polarized antennas at different frequencies is presented that gives circular polarized response. Thus, proposed work will be helpful to design similar circular polarized circular microstrip antenna at any desired frequency.

Keywords: Circular Microstrip Antenna, Circular Polarization, Rectangular Slot, Resonant Length Formulation

1. INTRODUCTION

The circular polarized (CP) microstrip antenna (MSA) is realized either by cutting narrow slot inside the coaxially fed MSA or by using dual feeds in conjunction with the power divider network, or by cutting asymmetrical U-slot or rectangular slot at an appropriate position inside the patch [1] -[11] . Amongst these techniques, U-slot or rectangular slot cut MSAs yields optimum response in terms of gain and axial ratio (AR) bandwidth (BW) [9] - [11] . However in these reported CP MSAs, formulation in resonant length at orthogonal modes and procedure to design them at any other frequency is not explained. Therefore, optimization process to design CP MSAs involves parametric study on different patch parameters using simulation software, followed by experimental validation. In this paper, first rectangular slot cut circular MSA (CMSA) to give CP response in 900 MHz frequency range, is discussed. On glass epoxy substrate (e_r = 4.3, h = 0.16 cm, tand = 0.02), it gives VSWR and AR BW of 20 and 7 MHz, respectively. To increase its gain, three layer suspended configuration is proposed which gives VSWR and AR BW of 32 and 9 MHz, respectively, with a gain of more than 3 dBi. Further by studying surface current distributions at orthogonal modes, formulation in their resonant length in terms of slot and patch dimensions for non-suspended and suspended MSAs, is proposed. The frequencies calculated using them closely agree with the simulated results. Using proposed formulations, procedure to design slot cut CP CMSA at any other frequency in non-suspended and suspended configuration is presented. It gives CP response with formation of small loop inside VSWR = 2 circle and with the broadside radiation pattern that shows co and cross polarization levels within 3 dB in E and H-planes. Thus, proposed formulations can be used to realize CP response at any desired frequency. All these MSAs were first designed and optimized using IE3D software [12] . For only the optimized and re-designed configurations, experiments were carried out to validate the simulated and proposed design procedure. The MSAs were fed using SMA panel type connector of 0.12 cm inner wire diameter. The impedance measurements were carried out using R & S vector network analyzer (ZVH-8). In simulations, infinite ground plane with infinite extent dielectric was assumed. This helps in realizing the simulation of MSA in reduced time. To realize this effect in measurements, size of substrate is taken to be more than six times the substrate thickness in all the direction with respect to patch periphery. The radiation pattern and the co-polar gain were measured using RF source (SMB 100A) and spectrum network analyzer (FSC6). The measurement was carried out in antenna lab wherein required minimum far field distance was kept between reference antenna and antenna under test.

2. SLOT CUT CP CMSA

The rectangular slot cut CMSA is shown in Figure 1(a). The units of the dimensions and frequencies referred in text and in figures are in cm and MHz, respectively. The CMSA radius "a" is calculated such that its TM_{11} mode frequency is 900 MHz [1] . To excite orthogonal modes, slot is cut in the patch center and coaxial feed is placed along the diagonal axis. The parametric study is carried out for length "l" = 0.6 ($0.02l_0$) and for increasing slot width (w). For two orthogonal modes, their frequencies and ratio plots against respective VSWR and AR BWs at each of slot width (w) are shown in Figure 1(b). The surface current distributions for "l" = 0.6 and w = 1.5 at two modes is shown in Figure 2(a) and Figure 2(b). The slot has degenerated TM_{11} mode into two orthogonal modes. At each mode, one half wavelength variation in surface currents is present along x and y-axis. The reduction in the frequency of first mode (f_1) is higher as slot width is orthogonal to its surface currents. The variation in second mode (f_2) frequency is marginal. Thus "f_2/f_1" ratio increases with increase in "w". The optimum result is obtained for w = 1.5 cm to give simulated VSWR and AR BW of 29 and 8 MHz, respectively, as shown in Figure 2(c) and Figure 2(d). The measured BW is 32 MHz. For w > 1.5 cm, f_2/f_1 ratio increases that do not maintain the required phase quadrature between two modes, which reduces AR BW.

As the slot cut CP CMSA is optimized on lossy substrate, the gain is less than 0 dBi. To improve the same, its three layer suspended configuration is proposed as shown in Figure 3(a) and Figure 3(b). The two layers of glass substrate are separated by an air gap of thickness 0.16 cm, thereby giving total substrate thickness of 0.48 cm. The patch is fabricated on top glass epoxy layer whereas SMA connector is connected to the bottom glass epoxy layer which acts as a ground plane. In suspended configuration, effective dielectric constant (e_{re}) reduces, that increases patch radius for f_{TM11} = 900 MHz. The e_{re} is calculated by using Equation (1) and further by using Equations (2) and (3), patch radius is calculated [1] . For 900 MHz, radius was found to be 6.5 cm. For suspended

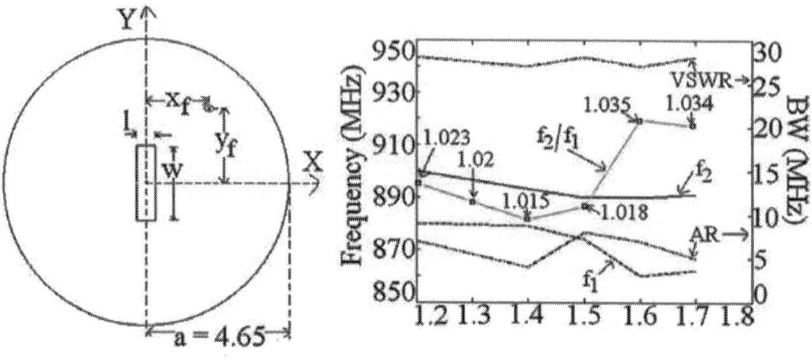

(a) (b) w (cm)

Figure 1. (a) Slot cut CP CMSA; its (b) dual frequency, ratio, AR and VSWR BW plots.

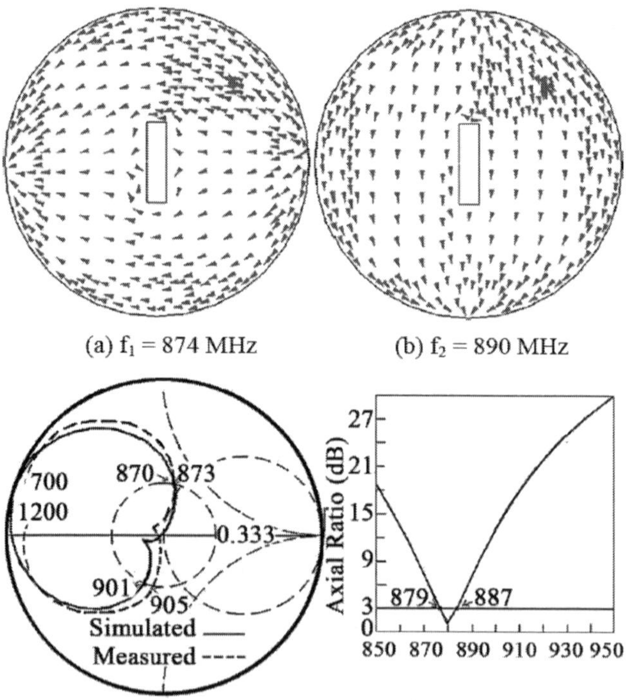

(c) (d) Frequency (MHz).

Figure 2. (a), (b) Surface current distribution at orthogonal modes and (c) input impedance and (b) simulated AR plots for slot cut CP CMSA.

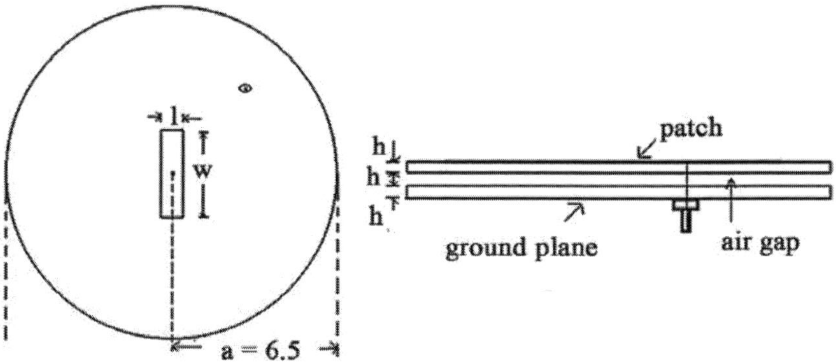

Figure 3. (a) Top and (b) side views of suspended slot cut CMSA.

slot cut CMSA, parametric study for variation in slot width is carried out and plots of orthogonal mode frequencies, their ratio, VSWR and AR BW's, are shown in Figure 4.

$$\varepsilon_{re} = \frac{3\varepsilon_r}{\varepsilon_r + 2} \quad (1)$$

$$k = \frac{8.794}{f_r\sqrt{\varepsilon_{re}}} \quad (2)$$

$$a = \frac{k}{\sqrt{1 + \frac{2h}{\pi\varepsilon_{re}k}\left(\ln\left(\frac{\pi k}{2h}\right) + 1.7726\right)}} \quad (3)$$

where,

h = glass epoxy substrate and air gap thickness, f_r = resonance frequency in GHz,

k = unit less number, a = patch radius in cm, e_r = substrate dielectric constant, e_{re} = effective dielectric constant.

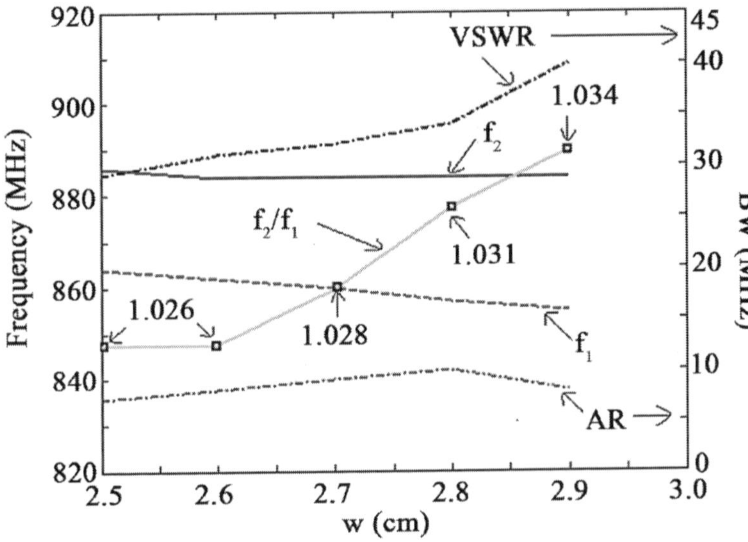

Figure 4. Dual frequencies, their ratio, VSWR and AR BW plots for suspended slot cut CP CMSA.

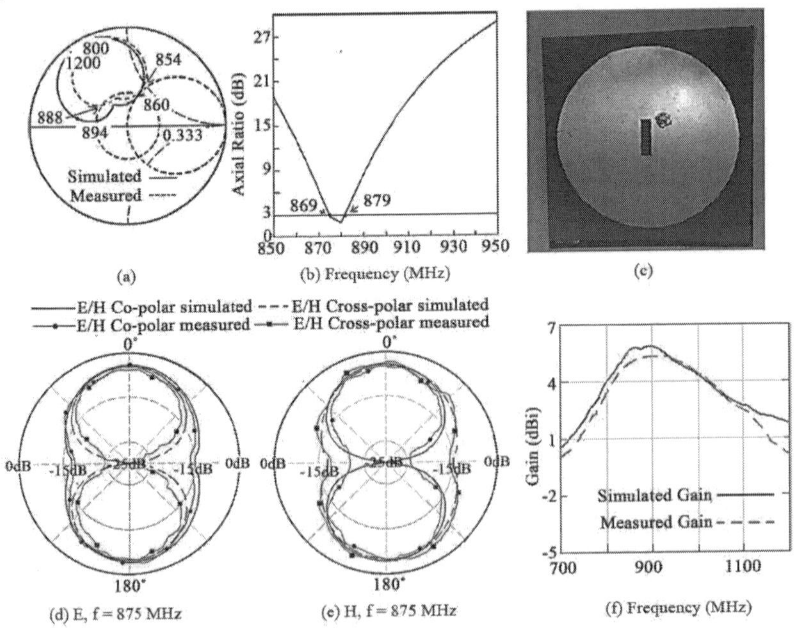

Figure 5. (a) Input impedance and (b) simulated AR plots, (c) fabricated prototype, (d), (e) radiation pattern and (f) gain variation over BW for suspended slot cut CP CMSA.

The optimum AR BW is obtained for w = 2.8 and results for the same are shown in Figure 5(a) and Figure 5(b). The simulated VSWR and AR BW's are 34 and 10 MHz, respectively. The suspended slot cut CP CMSA was fabricated and experiment was carried out. The measured BW is 32 MHz. The fabricated prototype of the configuration is shown in Figure 5(c). The surface current distribution at dual modes in suspended CMSA is similar to that present in non-suspended CMSA. The simulated and measured radiation pattern at center frequency of AR BW and gain variation over VSWR and AR BW's is shown in Figures 5(d)-(f). The co and cross polar levels are within 3 dB in the broadside direction which indicates the presence of CP. Due to use of suspended configuration antenna gain is more than 3 dBi. A good agreement is obtained between simulated and measured results. Although slot cut CP CMSAs discussed above are widely reported but the formulation to design them at any given frequency is not available. Hence most of the available work is based upon the parametric optimization. Here, by studying the surface current distributions, the formulation in resonant length at two orthogonal modes in slot cut CP CMSA is proposed. Further using proposed formulations, procedure to design slot cut CP CMSA at any other frequency is outlined.

3. FORMULATIONS FOR SLOT CUT CP CMSAS

In slot cut CMSA, surface currents at two modes are circulating around the slot length. At first mode, they are perturbed in their length due to slot width whereas at second mode, current length is modified due to slot length. The formulation in resonant length on glass epoxy substrate is obtained by modifying patch radius in terms of slot width or length as given in Equations (4) & (5). The frequencies at two modes are calculated by using resonance frequency equation for TM_{11} mode as given in Equations (6) and (7). Using Equation (8), % error between calculated and simulated values is obtained. The frequencies and % error plots are shown in Figure 6(a). At both the modes, closer agreement is obtained between simulated and calculated values. The formulation in resonant length for suspended CMSA as shown in Figure 3(a) and Figure 3(b), is obtained by modifying patch radius in terms of slot dimensions as given in Equations (9) and (10). The frequencies and % error are calculated by using Equations (6)-(8) and they are plotted in Figure 6(b). Over the entire range of slot dimension, closer agreement between simulated and calculated values is obtained.

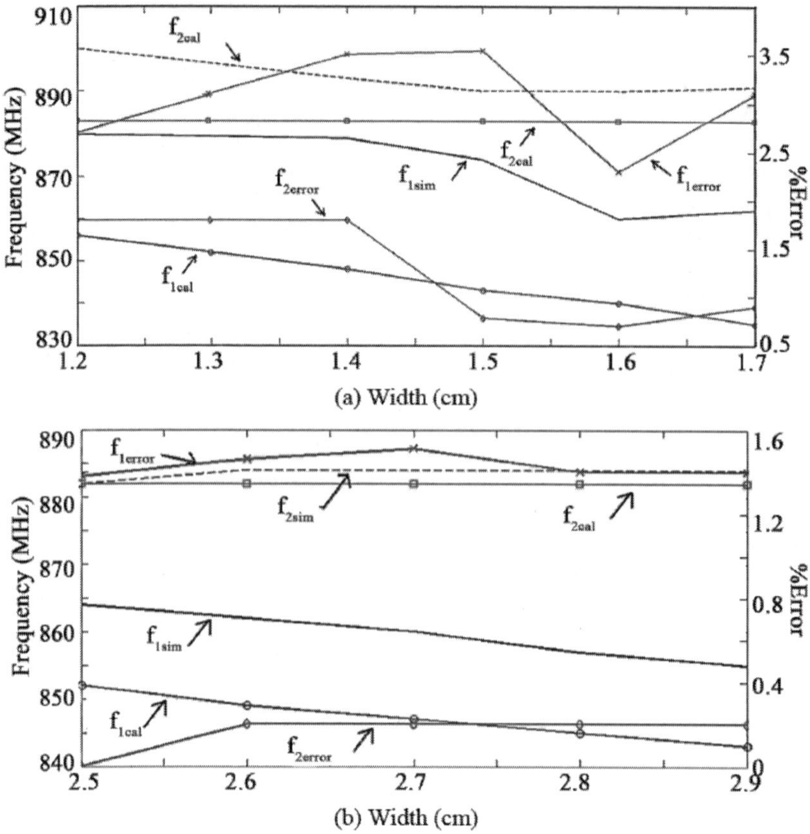

Figure 6. Dual frequency and % error plots for (a) non-suspended and (b) suspended slot cut CP CMSAs.

$$a_{e1} = a + \frac{w}{4} \tag{4}$$

$$a_{e2} = a + \frac{l}{4} \tag{5}$$

$$f_{r1} = \frac{1.84118c}{2a_{e1}\pi\sqrt{\varepsilon_r}} \tag{6}$$

$$f_{r2} = \frac{1.84118c}{2a_{e2}\pi\sqrt{\varepsilon_r}} \tag{7}$$

$$E = \left(\frac{f_{s1} - f_{r1}}{f_{s1}}\right) \times 100 \tag{8}$$

$$a_{e1} = a + \frac{w}{6} \tag{9}$$

$$a_{e2} = a + \frac{l}{6} \tag{10}$$

The substrate thickness at 900 MHz for non-suspended and suspended configuration is $0.005l_0$ and $0.014l_0$, respectively. Since difference in substrate thickness in two configurations is not larger, their resonant length formulations are nearly identical. As seen from Figure 6(a) and Figure 6(b), calculated frequencies show 2% to 3% error with respect to simulated frequencies. As seen from Figure 1(b) and Figure 4, in optimized slot cut CP CMSAs, simulated f_2/f_1 ratio is 1.02 and 1.03 for non-suspended and suspended MSAs, respectively. The plots of simulated and calculated dual frequencies and their ratios are shown in Figure 7(a) and Figure 7(b). For the optimized slot width, calculated f_2/f_1 ratio nearly equals 1.05 and 1.045 in non-suspended and suspended CMSAs, respectively. Using the proposed formulations, procedure to design slot cut CP CMSAs at different frequencies is presented below.

4. DESIGN PROCEDURE FOR SLOT CUT CP CMSAS

In slot cut CP CMSAs at 900 MHz, slot length (l) is $0.02l_0$ whereas feed point is placed along the diagonal axis, i.e. at $x_f = y_f = 0.06l_0$. Using proposed formulations, design of slot cut CP CMSA is presented for equivalent patch TM_{11} mode frequency of 700, 1500, 2000 and 3000 MHz, on glass epoxy substrate. The patch radius is calculated by using equations (2) and (3) and at respective frequencies it is found to be, 6.0, 2.8, 2.05 and 1.33 cm. The plots of orthogonal mode frequencies and their ratio are generated by using proposed formulations for $l = 0.02l_0$ and increasing slot width, and they are shown in Figures 8(a)-(d). From these plots, value of slot width that gives

f_2/f_1 of approximately 1.05, is selected. At 700, 1500, 2000 and 3000 MHz, slot width is found to be 2.0, 1.0, 0.7 and 0.6 cm, respectively.

At respective frequencies, slot of selected width and $0.02l_0$ in length is cut in the patch center with feed is placed at $x_f = y_f = 0.06l_0$, along the diagonal axis. The slot cut CMSAs are simulated and their input impedance plots are shown in Figure 9(a) and Figure 9(b) and Figure 10(a) and Figure 10(b).

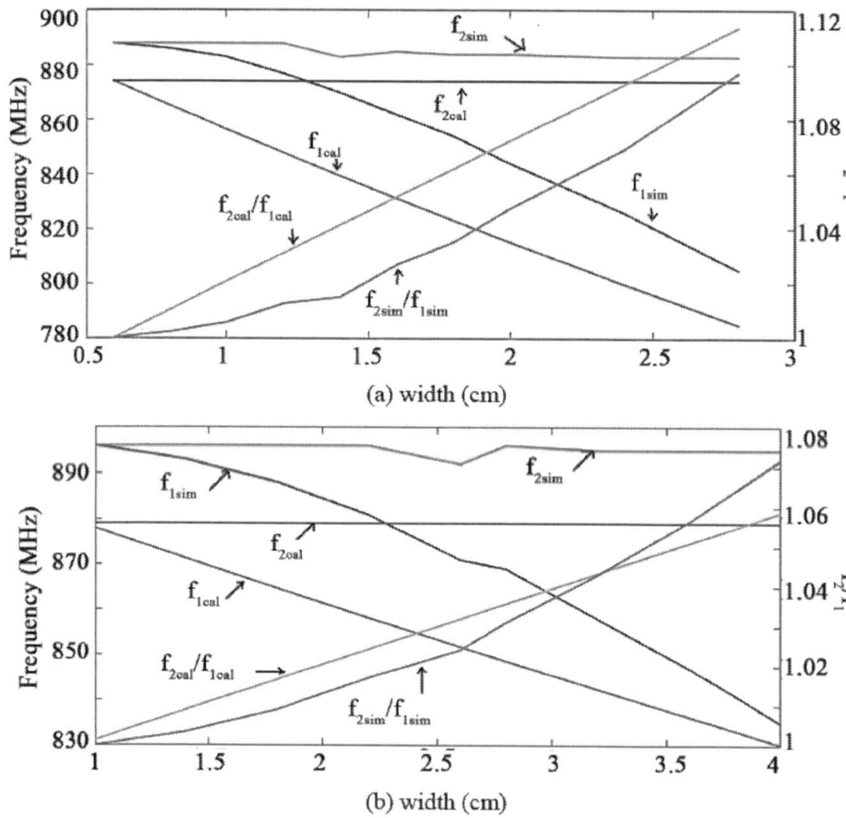

Figure 7. Simulated and calculated dual frequencies and ratio plots for (a) non-suspended and (b) suspended CP CMSAs.

CMSA and (c), (d) calculated dual frequencies and ratio plots for suspended slot cut CP CMSAs.

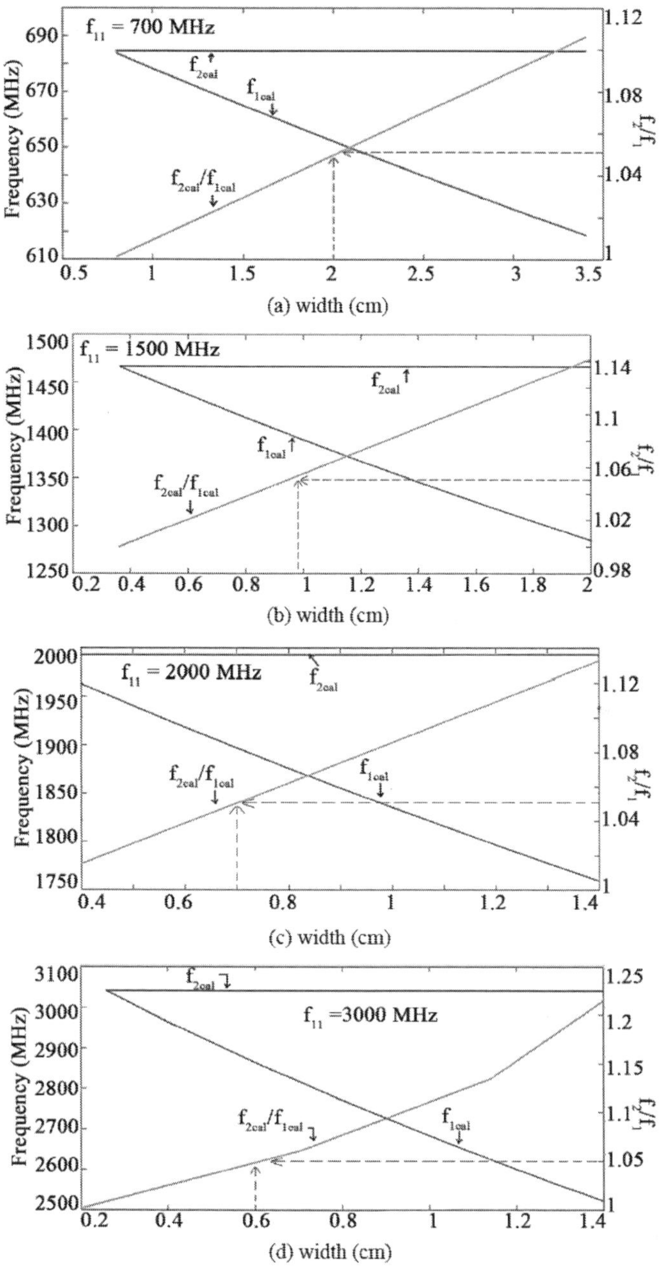

Figure 8. (a)-(d) Dual frequencies and their ratio plots for slot cut CMSA at different frequencies.

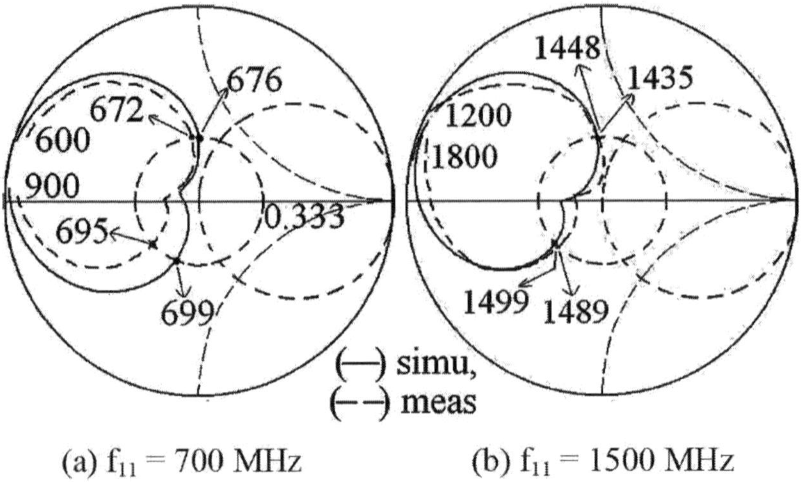

Figure 9. Input impedance plots at (a) 700 and (b) 1500 MHz for slot cut CP CMSA.

At all the frequencies CP response with formation of small loop (kink) inside VSWR = 2 circle is obtained. The simulated VSWR and AR BW at 700 and 1500 MHz are, 23 and 6 MHz and 54 and 13 MHz, respectively. The simulated VSWR and AR BW's at 2000 and 3000 MHz are, 71 and 18 MHz and 131 and 26 MHz, respectively. The measured BW at 700, 1500, 2000 and 3000 MHz are, 23, 51, 73 and 131 MHz, respectively. At each of the frequencies, simulated and measured radiation pattern, in the broadside direction shows co and cross polarization levels within 3 dB, which indicates the presence of CP. Further suspended slot cut CP CMSAs are designed. To realize impedance matching for coaxially fed suspended MSAs, at each of the frequencies, the effective substrate thickness is selected to be $<0.03l_0$. Thus at 700, 1500 and 2500 MHz, three layer suspended configuration with an air gap of 0.16 cm between two glass epoxy layer is selected. At 3000 MHz, two layer suspended configuration in which glass epoxy layer suspended above the ground plane is used. The patch radius is calculated by using Equations (1)-(3) and it is found to be, 8.3, 3.6, 2.6 and 1.9 cm at 700, 1500, 2000 and 3000 MHz, respectively. Using proposed formulations for suspended CP CMSAs, plots of orthogonal frequencies and their ratio are generated for $l = 0.02l_0$ and increasing slot width. At each of the frequencies they are shown in Figure 10(c) and Figure 10(d) and Figure 11(a) and Figure 11(b).

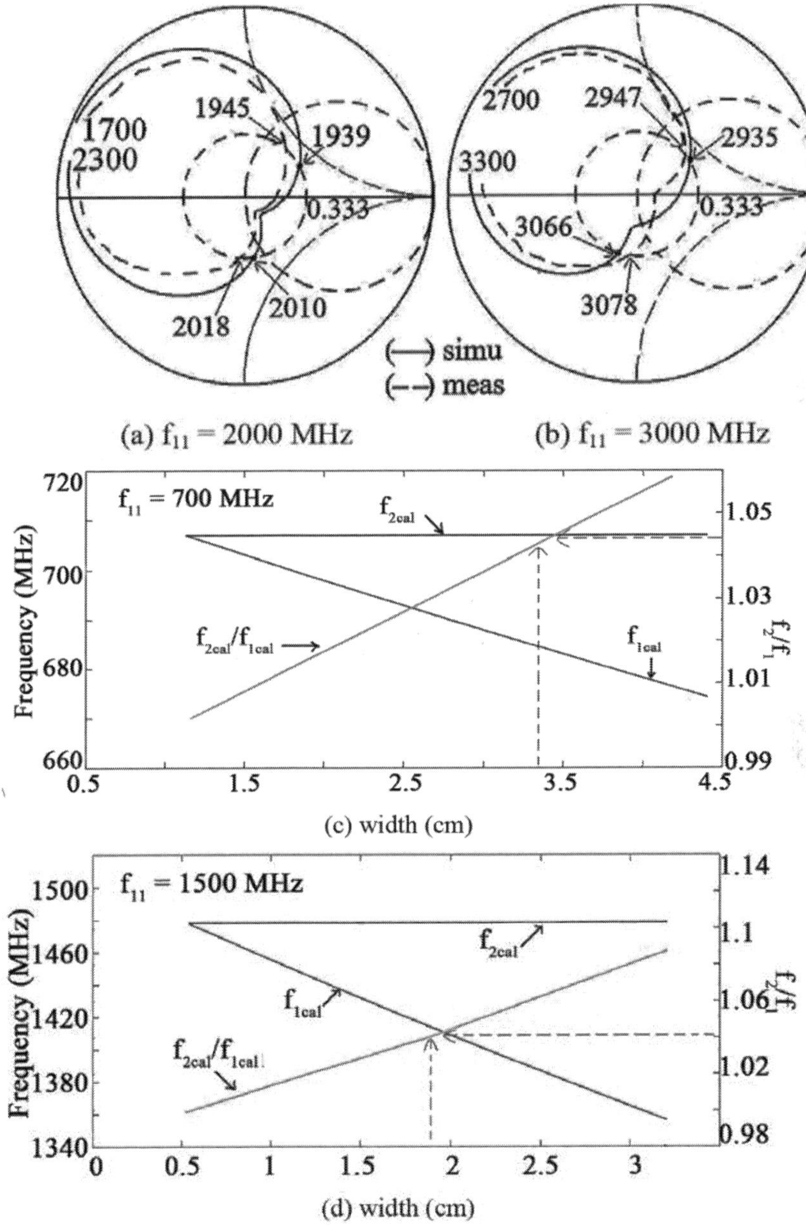

Figure 10. (a), (b) Input impedance plots at two frequencies for slot cut CP

The value of slot width that gives f_2/f_1 of nearly 1.045 is selected as shown in their respective plots. At 700, 1500, 2000 and 3000 MHz, slot width was found to be, 3.4, 1.8, 1.6 and 1.1 cm, respectively. The slot of $0.02l_0$ in length (l) and selected slot width is cut inside suspended CMSAs. The simulated and measured input impedance plots for slot cut CMSAs are shown in Figures 12(a)-(d). At all the frequencies, small loop is observed inside VSWR = 2 circle in the smith chart. The simulated VSWR and AR BW at 700 and 1500 MHz are, 21 and 8

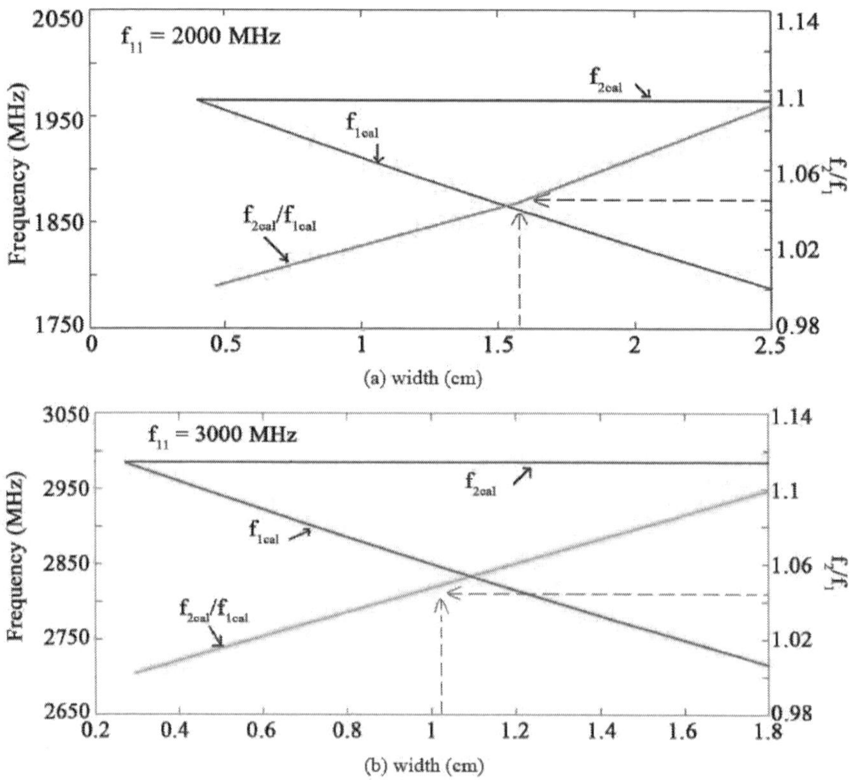

Figure 11. (a), (b) Calculated dual frequencies and their ratio plots for suspended slot cut CP CMSAs at different TM11 mode frequencies.

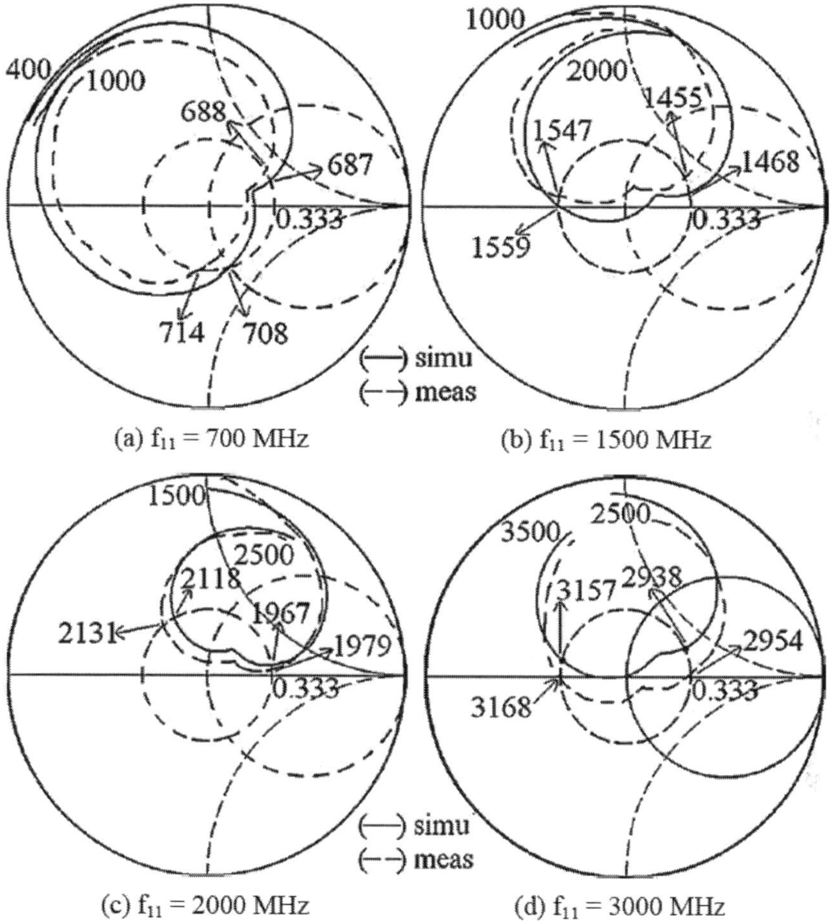

Figure 12. (a)-(d) Input impedance plots for suspended slot cut CP CMSAs at different TM11 mode frequencies.

MHz and 91 and 17 MHz, respectively. The simulated VSWR and AR BW's at 2000 and 3000 MHz are, 151 and 18 MHz and 219 and 26 MHz, respectively. The measured BW at 700, 1500, 2000 and 3000 MHz are, 26, 92, 152 and 214 MHz, respectively. The simulated and measured pattern plots at center frequency of AR BW for suspended CP CMSAs at 1500 and 3000 MHz is shown in Figures 13 (a)-(d). The pattern shows co and cross levels within 3 dB, which indicates the presence of CP. For f_{TM11} = 900 MHz, using the proposed formulations slot cut CP CMSA is also designed on RT-duroid substrate (e_r = 2.33, h = 0.16 cm, tand = 0.001). The patch radius for non-suspended and suspended configurations was calculated to be 6.5 and 7.5 cm, respectively. Using proposed formulations, plots for dual frequencies and their ratio are generated as shown in Figure 14(a) and Figure 14(b). For frequency ratio of 1.05 and 1.045 for non-suspended and suspended CMSAs, slot width was found to be 1.5 and 2.4 cm, respectively. The slot of $0.02l_0$ in length and selected slot width is cut inside the circular patch. The simulated and measured input impedance plots are shown in Figure 15(a) and Figure 15(b).

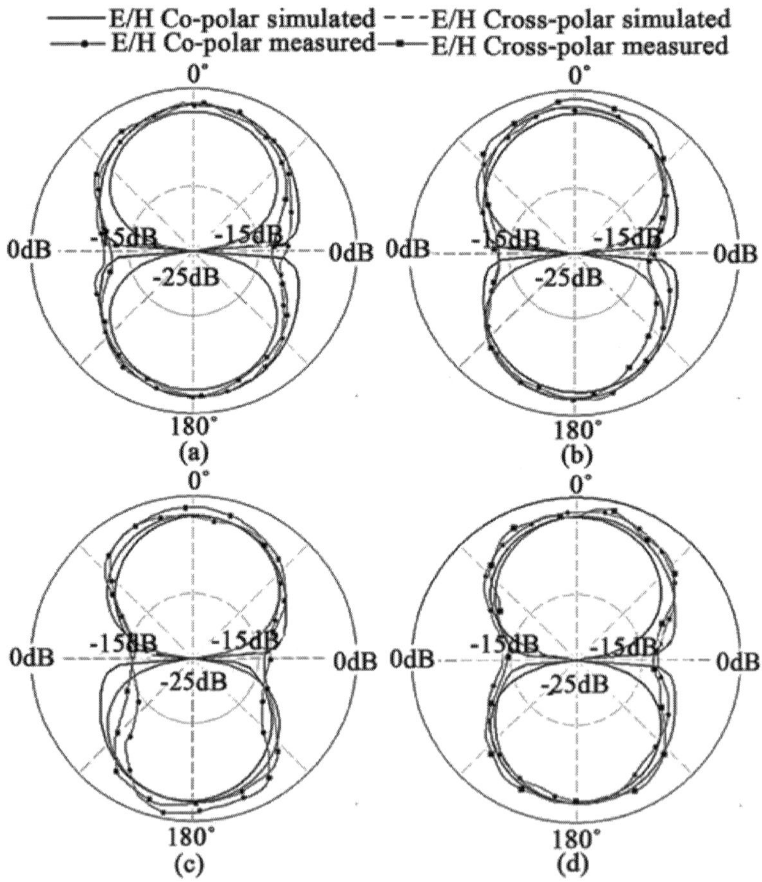

Figure 13. Radiation pattern at center frequency of AR BW at ((a), (b)) 1500 and ((c), (d)) 3000 MHz for suspended slot cut CP CMSAs. (a) (b) E, f = 1500 MHz, (c) (d) E, f = 3000 MHz.

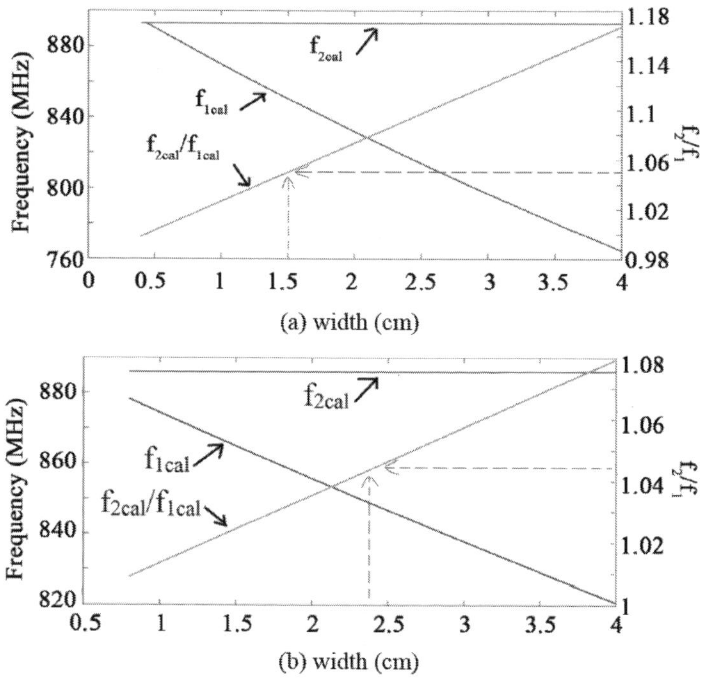

Figure 14. Dual frequencies and their ratio plots for (a) non-suspended and (b) suspended CMSAs on RT-duroid substrate for f_{TM11} = 900 MHz.

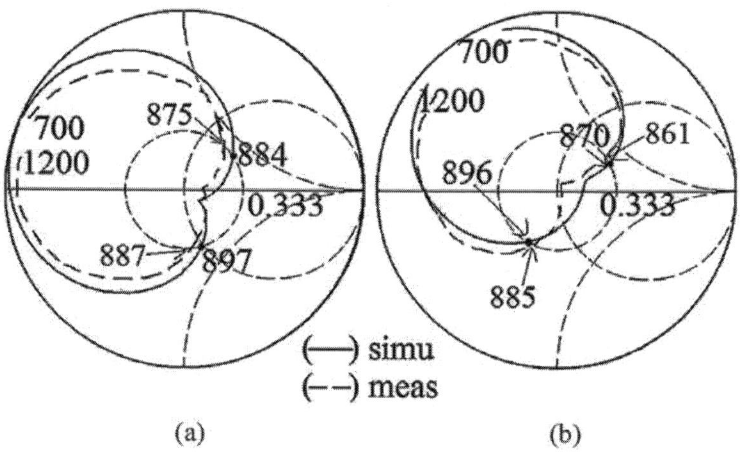

Figure 15. Input impedance plots for (a) non-suspended and (b) suspended CMSAs on RT- duroid substrate for f_{TM11} = 900 MHz.

For non-suspended CMSA, simulated VSWR and AR BWs are, 13 and 5 MHz, respectively. The measured BW is 12 MHz. For suspended CMSAs simulated values are, 26 and 4 MHz, respectively whereas measured BW is 24 MHz. The pattern shows difference between co and cross polarization levels to be less than 3 dB in the broadside direction. Thus proposed work will be useful for designing slot cut CP CMSAs at any desired frequency and on different substrates.

5. CONCLUSION

The design of narrow slot cut CP CMSA at 900 MHz is discussed. The slot degenerates patch TM_{11} mode into two orthogonal modes, to realize CP response. To improve the gain, three layer suspended design of slot cut CP CMSA is also proposed which gives gain of more than 3 dBi. In the reported literature on slot cut CP CMSAs, design guidelines to realize similar CP antennas at any given frequency are not available. Therefore by studying the surface current distributions at two orthogonal modes, formulation in resonant length in terms of slot and patch dimensions for non-suspended and suspended configurations, is proposed. The frequencies calculated using them closely agree with the simulated results. Further using proposed formulations, procedure to design similar slot cut CP CMSAs at different frequencies and on different substrates, is presented. At respective frequencies, it gives CP response with formation of small loop inside VSWR = 2 circle. Thus, proposed formulation and design procedure will be helpful for designing similar slot cut CP patch antennas. The proposed work can also provide some guideline to realize similar design procedure for other slot cut variations of CMSAs, rectangular MSA, equilateral triangular MSA and other modified patch shapes.

REFERENCES

1. Kumar, G. and Ray, K.P. (2003) Broadband Microstrip Antennas. 1st Edition, Artech House, Norwood.
2. Garg, R., Bhartia, P., Bahl, I. and Ittipiboon, A. (2001) Microstrip Antenna Design Handbook. 1st Edition, Artech House, Norwood.
3. Bhartia, B. and Bahl, I.J. (1980) Microstrip Antennas, USA.

4. Nasimuddin, Chen, Z.N. and Quing, X. (2012) A Compact Circularly Polarized Cross-Shaped Slotted Microstrip Antenna. IEEE Transactions on Antennas and Propagation, 60, 1584-1588.
5. Yang, S.L.S., Lee, K.F. and Kishk, A.A. (2008) Design and Study of Wideband Single Feed Circularly Polarized Microstrip Antennas. Progress in Electronics Research, 80, 45-61.
6. Kasabegoudar, V.G. and Vinoy, K.J. (2009) Design and Study of Broadband Suspended Microstrip Antenna for Circular Polarization. Progress in Electronics Research, 90, 353-368.
7. Deng, J., Guo, L., Fan, T., Wu, Z., Hu, Y. and Yang, J. (2013) Wideband Circularly Polarized Suspended Patch Antenna with Indented Edge and Gap-Coupled Feed. Progress in Electronics Research, 135, 151-159.
8. Sharma, P.C. and Gupta, K.C. (1983) Analysis and Optimized Design of Single Feed Circularly Polarized Microstrip Antennas. IEEE Trans Antennas and Propagation, 31, 949-955.
9. Lam, K. Y., Luk, K. M., Lee, K. F., Wong, H., and Bong, K. (2011) Small Circularly Polarized U-Slot Wideband Patch Antenna. IEEE Antennas and Wireless Propagation Letters, 10, 87-90.
10. Khidre, A., Lee, K.F., Yang, F. and Eisherbeni, A. (2010) Wideband Circularly Polarized E-Shaped Patch Antenna for Wireless Applications. IEEE Antennas and Propagation Magazine, 52, 219-229.
11. Tong, K.F. and Wong, T.P. (2007) Circularly Polarized U-Slot Antenna. IEEE Transactions on Antennas and Propagation, 55, 2382-2385.
12. IE3D 12.1, Zeland Software, Fremont, 2004.

4

CHAPTER

PATTERN SWITCHABLE ANTENNA SYSTEM USING INKJET-PRINTED DIRECTIONAL BOW-TIE FOR BI-DIRECTION SENSING APPLICATIONS

Seung-Hyun Eom, Yunsik Seo and Sungjoon Lim *

Electrical and Electronics Engineering, Chung-Ang University, 84 Heukseok-ro, Dongjak-gu, Seoul 06974, Korea

ABSTRACT

In this paper, we propose a paper-based pattern switchable antenna system using inkjet-printing technology for bi-direction sensor applications. The proposed antenna system is composed of two directional bow-tie antennas and a switching network. The switching network consists of a single-pole-double-throw (SPDT) switch and a balun element. A double-sided parallel-strip line (DSPSL) is employed to convert the unbalanced microstrip mode to the balanced strip mode. Two directional bow-tie antennas have different radiation patterns because of the different orientation of the reflectors and antennas. It is demonstrated from electromagnetic (EM) simulation and measurement that the radiation patterns of the proposed antenna are successfully switched by the SPDT switch.

Keywords: bow-tie antenna; inkjet-printing technology; paper-based antenna; pattern switchable antenna; reconfigurable antenna

INTRODUCTION

Reconfigurable antennas have attracted attention because of their potential applications and advantages. There are many types of reconfigurable antennas: radiation pattern, frequency, polarization, and combined antennas. Especially, radiation pattern reconfigurable antennas can overcome noisy environments, improve security, prevent from electronic jamming, and save energy [1]. Up to now, pattern reconfigurable antennas have been realized using various switching component such as PIN diodes, varactor diodes, MEMS (micro-electro-mechanical systems) switches, and FET (field effect transistor) components [2–6].

In addition, there is a growing interest in easy-to-use inkjet-printing technology to manufacture electronics on flexible substrates. Inkjet printing is a noncontact printing technology. It becomes possible to print patterns on a substrate directly with the droplets ejected from the nozzle. Compared with inkjet-printing technology, conventional photolithography technology has a high exposure apparatus cost, a complex process, environmental pollution, and large material losses that accompany the process [7]. A direct printing process such as inkjet-printing technology consists only of the printing process and the sintering process. Therefore, a complicated and long-duration process can be replaced by a cost-effective and simple process. With inkjet-printing technology, there is no unnecessary loss of materials. Flexible electronics can be inkjet-printed on flexible substrates such as polyethylene, terephthalate, polyimide, and paper [8–10]. Inkjet-printing technology has been used in various RF applications such as the inkjet-printed radio-frequency identification (RFID) tag [11], ultrawideband antennas [12], gas-detection sensor [13], high frequency inductors and capacitors [14], frequency selective surfaces (FSS) [15], composite right/left-handed transmission line [16], flexible metamaterial absorber [17], terahertz split ring resonator [18] and artificial magnetic conductor (AMC) structure [19].

In this study, a paper-based pattern switchable antenna is proposed using inkjet-printing technology. The proposed antenna system is composed of two directional bow-tie antennas and a switching network. The switching network is built on a printed-circuit-board (PCB) and consists of a balun element and an SPDT switch. The bow-tie antennas with two different radiation patterns are printed on paper, and a reflector is added on the same paper to enhance the directivity. Because of two different radiation patters, the proposed antenna system is useful for bi-direction sensor applications. The final antenna system is realized by bonding the inkjet-printed antennas on paper and the switching network on the PCB. In order to achieve low-cost fabrication, we used a commercial inkjet printer that is commonly used at home, instead of an expensive material printer that is usually used in industry.

ANTENNA SYSTEM DESIGN

Directional Bow-Tie Antenna Design

Figure 1 illustrates two directional bow-tie antennas. First, an omnidirectional bow-tie antenna is designed to resonate at 1.8 GHz. Its parameters are determined by Equation (1)–(5) [20,21]:

$$f_r = \frac{c}{2\sqrt{\varepsilon_e}}\left(\frac{1.152}{R_t}\right) \tag{1}$$

$$R_t = \frac{L}{2}\frac{(W+2\Delta l)+(W_c+2\Delta l)}{(W+2\Delta l)(S+2\Delta l)} \tag{2}$$

$$\Delta l = h\frac{0.412(\varepsilon_e+0.3)(\frac{W_i}{h}+0.262)}{(\varepsilon_e-0.258)(\frac{W_i}{h}+0.813)} \tag{3}$$

$$\varepsilon_e = \left(\frac{\varepsilon_r+1}{2}\right)+\left(\frac{\varepsilon_r-1}{2}\right)\left(1+\frac{12h}{W_i}\right)^{-1/2} \tag{4}$$

$$W_i = \left(\frac{W+W_c}{2}\right) \tag{5}$$

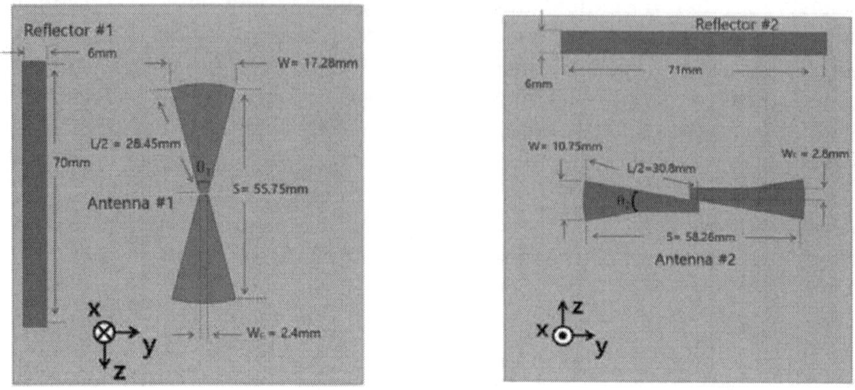

(a) (b)

Figure 1. Layout of two inkjet-printed bow-tie antennas. (**a**) first antenna and (**b**) second antenna.

Table 1. Dimension of Bow-Tie Antennas.

	W	S	L/2	W_c
Antenna #1	17.28 mm	55.75 mm	28.45 mm	2.4 mm
Antenna #2	10.75 mm	58.26 mm	30.8 mm	2.8 mm

The relative permittivity, effective permittivity, and thickness of the substrate are denoted by εr, εe, and h. The other geometrical parameters are defined in Figure 1 and Table 1. For pattern-switching capability, a directional antenna is preferred. Because a bow-tie antenna has an omnidirectional radiation pattern, a directional bow-tie antenna is designed by loading a reflector as shown in Figure 1 [22,23]. In addition, two directional bow-tie antennas are arranged orthogonally to each other for high isolation. We determined the optimum position, length, and width of the reflector to achieve highest directivity. Its performances are simulated by ANSYS high frequency structure simulator (HFSS). Figure 2 shows the simulated S-parameters of the antennas #1 and #2. At 1.8 GHz, the simulated return losses of the antenna #1 and #2 are 19.62

dB and 38.99 dB, respectively. The isolation between two antennas (S21) is ´13.49 dB at 1.8 GHz. The simulated 10-dB impedance bandwidths of the antenna #1 and #2 are 1.69–2.07 GHz and 1.69–2.19 GHz, respectively. The bandwidth of each antenna is controlled by the width Wc [24].

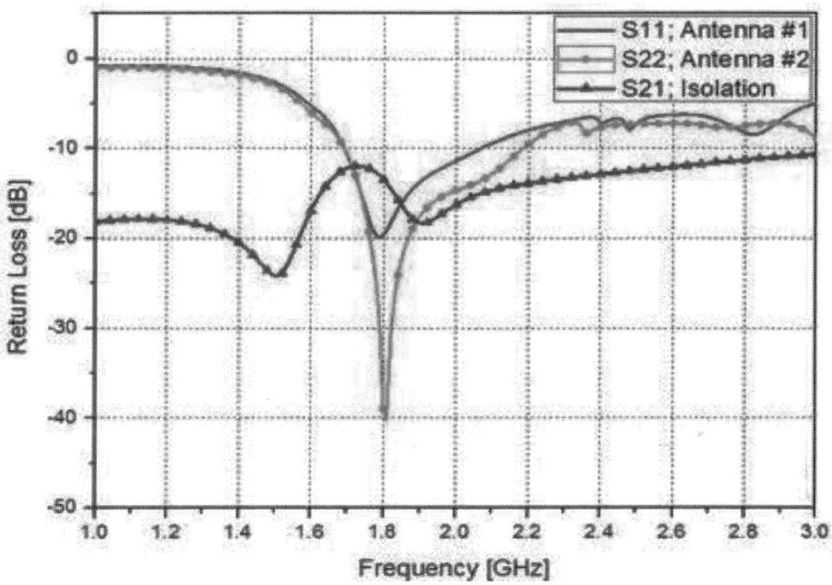

Figure 2. Simulated S-parameters of the designed antennas; return loss of the first antenna (S11), return loss of the second antenna (S22), and isolation between two antennas (S21).

Figure 3a,b show the simulated 3D radiation patterns of the antenna #1 without and with the reflector, respectively. Due to the reflector, the peak gain is increased from 1.85 dBi to 4.47 dBi. Figure 3c,d show the simulated 3D radiation patterns of the antenna #2 without and with the reflector, respectively. Due to the reflector, the peak gain is increased from 2.29 dBi to 5.15 dBi. As shown in Figure 3b,d, the maximum radiation direction of the antenna #1 and #2 are orthogonal each other. Therefore, it is expected that the radiation pattern can be switched by selecting one of two antennas. The radiation efficiencies of the antenna #1 and the antenna #2 are 88.49% and 87.5%, respectively.

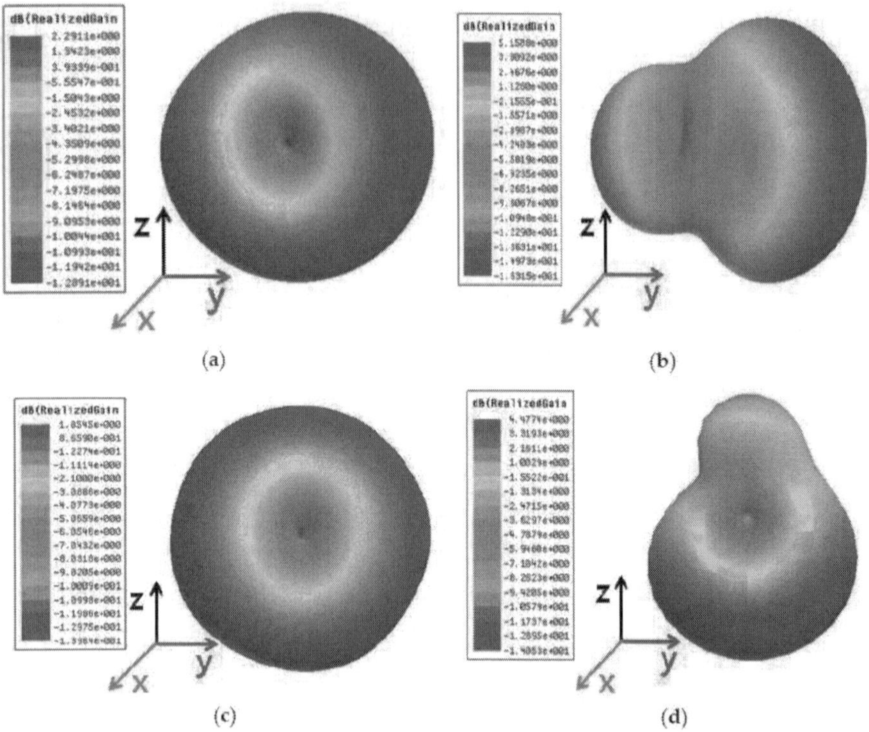

Figure 3. Simulated 3D radiation patterns when (**a**) first antenna is on without reflector, (**b**) first antenna is on with reflector, (**c**) second antenna is on without reflector and (**d**) second antenna is on with reflector.

Switching Network Design

In order to select one of two directional antennas, the switching network is designed. Because the bow-tie antennas are fed by a balanced signal, a balun is required in the switching network. In this work, a double-sided parallel-strip line (DSPSL) is introduced to transform from unbalanced signal of the microstrip line to balanced signal.

The geometrical parameters of the switching network which consists of the DSPSL balun and SPDT switch are indicated in Figure 4. The 50-ohm microstrip line width is 2.35 mm, and the DSPSL line width is 2.4 mm. When the height of the substrate is the same, the DSPSL has a higher characteristic impedance than the microstrip line. Therefore, the DSPSL line width is greater than the microstrip line width.

In this work, AS193-73LF (Skyworks Solution, Inc.) is used as the SPDT switch which select the antenna #1 or #2. Figure 4c shows the layout of the SPDT switch with the bias network where three DC blocking capacitors with 100 pF are used at PIN2, PIN4, and PIN6. When a 5 V DC is connected to PIN1, as a result PIN2 is connected to PIN4. Similarly, when a 5 V DC is applied to PIN3, PIN2 is connected to PIN6. The insertion loss and isolation of the SPDT switch are typically 0.45 dB and 19 dB at 1.0–2.0 GHz. The return loss of the SPDT switch is typically 17.7 dB at 1.0–2.5 GHz. For full-wave simulation, the SPDT switch is represented by an equivalent circuit model. In the ON state, the SPDT switch is represented as a series resistance (R_S = 1.5 ohm) and a series parasitic inductance (L_S = 0.02 nH), In the OFF state, the SPDT switch is represented as a shunt capacitance (C_t = 0.015 pF) with a shunt resistance (R_p) and a series parasitic inductance (L_S = 0.02 nH). However, R_p of the OFF state can be neglected because resistance is greater than the reactance of C_t.

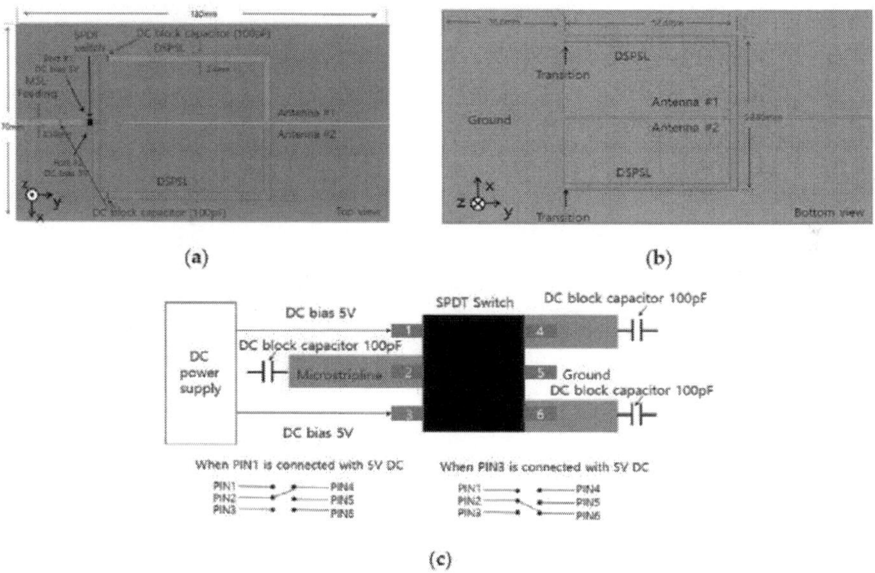

Figure 4. Layout of switching network; (**a**) top view and (**b**) bottom view (**c**) single-pole-double-throw (SPDT) switch with bias network.

Figure 5a shows a three-dimensional view of the proposed antenna with the switching network. As shown in the inset of Figure 5a, the top line of the DSPSL #2 is connected to the left arm of antenna #2 and the bottom line of the DSPSL is connected to the right arm of the antenna #2. Similarly, the top line of the DSPSL #1 is connected to the left arm of antenna #1 and the bottom line of the DSPSL #1 is connected to the right arm of the antenna #1. A microstrip line is an unbalanced line and a DSPSL is a balanced line. Figure 5b,c shows the electric field distributions of the microstrip line and DSPSL, respectively. When the perfect electric conductor is placed at the center of the substrate and parallel to the strips, the DSPSL is identical to two back-to-back placed microstrip lines. Thus, it can be easily analyzed using image theory [25–27]. At the DSPSL output (B-B^1 plane), out-of-phase signals are combined while the in-phase signals are cancelled out. In addition, the DSPSL has wider linewidth compared to a microstrip line with the same characteristic impedance because the guided wavelength of the DSPSL is lower than that of the microstrip line.

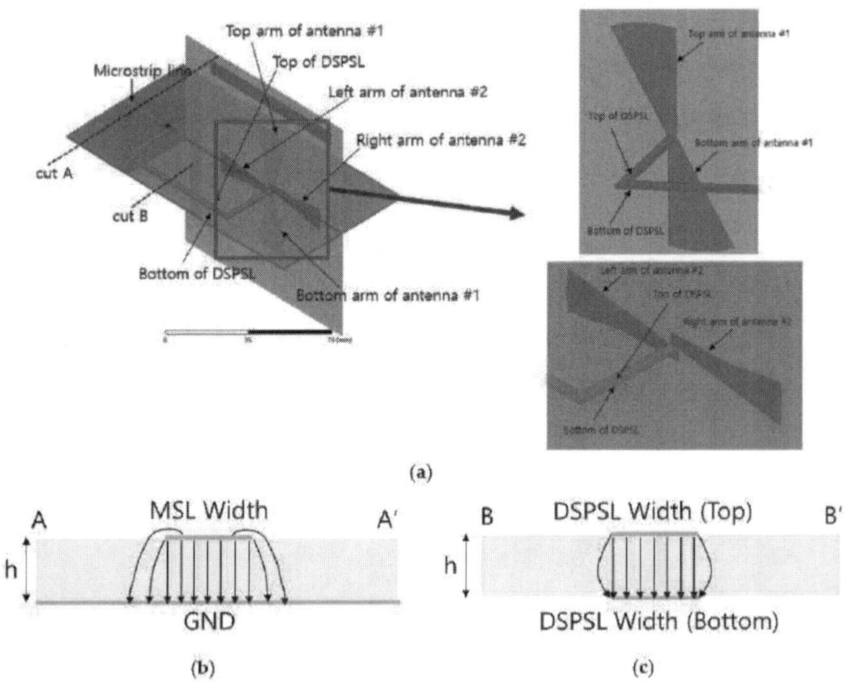

Figure 5. (**a**) 3D-view of antenna with switching network; (**b**) E-field distribution of cut A plane (microstrip line); (**c**) E-field distribution of cut B plane (double-sided parallel-strip transmission line).

FABRICATION AND MEASUREMENTS

Two directional bow-tie antennas are inkjet-printed on photo paper using a home printer (Epson WF-7011) which is shown in Figure 6a. We used silver nanoparticle ink (JS-B25P, Novacentrix) for printing which contains 25% silver as shown in Figure 6b. The sheet resistance of the silver nanoparticle ink is 60 mΩ. To increase the conductivity, sintering process is necessary. At lower temperature, large interval exists between the particles. At higher temperature, gap of the particles start to diminish [11,22]. The photo paper can withstand a temperature of 180 ″C. Therefore, a sintering process at a temperature of 180 ″C was used.

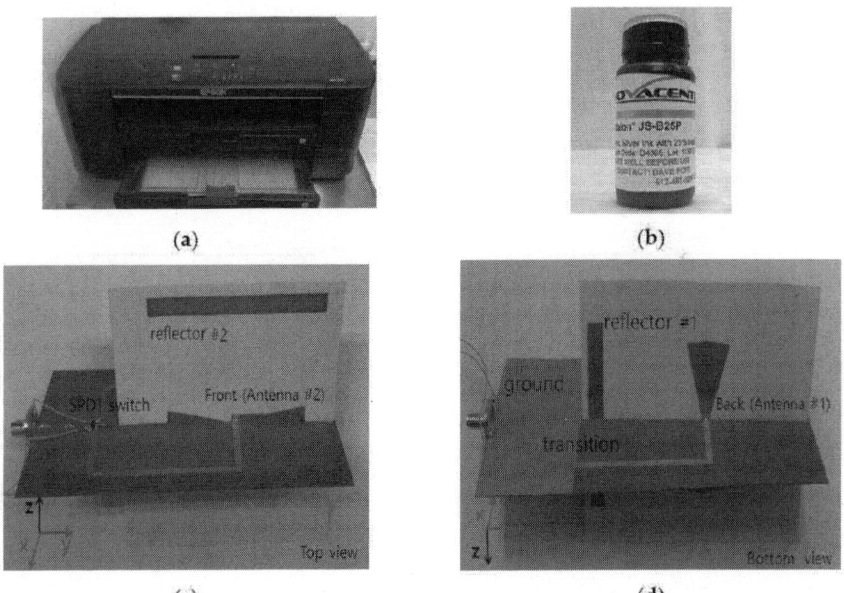

Figure 6. Pictures of home inkjet printer, silver nanoparticle ink and fabricated antenna system; home printer (WF-7011); (**b**) silver ink (JS-B25); (**c**) prototype antenna top view; and (**d**) prototype antenna bottom view.

The proposed bow-tie antennas are inkjet-printed separately, one on each of the two photo papers. Next, two inkjet-printed bow-tie antennas are bonded together in such a way that one antenna remains on the top and the other remains on the bottom side. A nonconductive adhesive material is used to bond them. The switching network was fabricated on a substrate of Duroid-5870 having a dielectric constant of 2.33, a loss tangent of 0.0012, and a thickness of 0.78 mm.

Nonconductive epoxy is used to bond the switching network on the Duroid-5870 and the inkjet-printed antennas on photo paper. Conductive silver epoxy (CW2400) is used to bond the signal lines. The volume resistivity of CW2400 is less than 0.001 ohm-cm; therefore, it has high enough conductivity to transmit signals. Figure 6c,d show the pictures of top and bottom view of the fabricated antenna system prototype.

A vector network analyzer was used for measurement of the fabricated antenna system. Figure 7 shows simulated and measured return loss of the proposed antenna system. When the antenna #1 is selected, the measured 10 dB impedance bandwidth is 1.61–2.22 GHz. When the antenna #2 is selected, the measured 10 dB impedance bandwidth is 1.66–2.2 GHz. Both modes provide a 10 dB impedance bandwidth of approximately 600 MHz. It can be confirmed that look similar results compared to the simulated results. When the antenna #2 is on, the resonance frequency was considerably shifted.

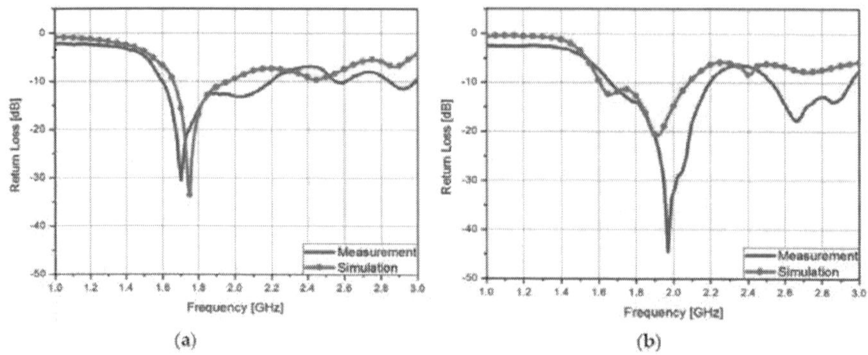

Figure 7. Simulated and measured return losses of the proposed antenna system when **(a)** first antenna is on and **(b)** second antenna is on.

The radiation pattern of the proposed antenna system is measured in an anechoic chamber. Figure 8 shows the normalized results of the radiation patterns obtained by the measurement when antenna #1 is on and antenna #2 is off, and vice versa. It is successfully demonstrated that the maximum radiation direction can be switched. The measured peak gains are 3.92 dBi and 4.72 dBi when antenna #1 and antenna #2 are selected, respectively.

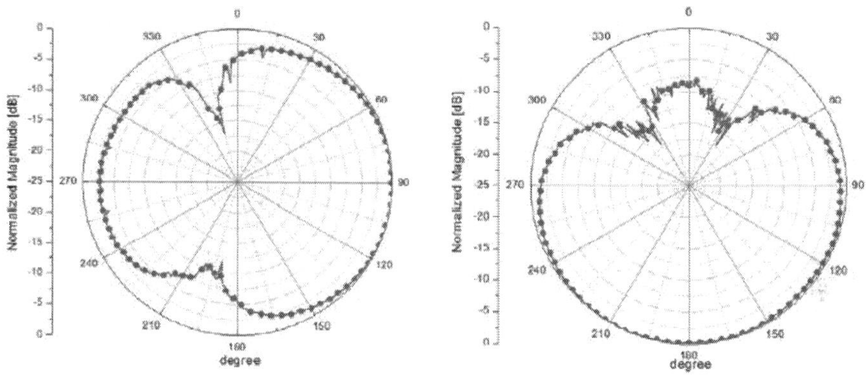

(a) (b)

Figure 8. Measured E-plane radiation patterns of fabricated antenna when (**a**) first antenna is on and (**b**) second antenna is on.

CONCLUSIONS

In this paper, we proposed a paper-based pattern switchable antenna using inkjet technology. Two directional bow-tie antennas with different radiation patterns are inkjet-printed on paper. In both radiation modes, the 10 dB impedance bandwidth is in the range 1.66–2.2 GHz. When antenna #1 and antenna #2 are selected, the peak gains are 3.96 and 4.72 dBi, respectively. It is numerically and experimentally demonstrated that two different radiation patterns are successfully switched by using the SPDT switch. Therefore, the proposed pattern switchable antenna can sense the objects at different location. In the present prototype, the switching network is fabricated on the PCB and the antennas are inkjet-printed on paper. This hybrid structure requires mechanical assembly processes. As future works, all-inkjet-printed prototype will be built by inkjet-printing the switching network on paper.

ACKNOWLEDGMENTS

This research was supported by the MSIP (Ministry of Science, ICT and Future Planning), Korea, under the ITRC (Information Technology Research Center) support program (IITP-2015-H8501-15-1007) supervised by the IITP (Institute for Information & communications Technology Promotion) and the National Research Foundation of Korea (NRF) grant funded by the Korea government (MSIP) (No. 2014R1A2A1A11050010).

AUTHOR CONTRIBUTIONS

Seung-Hyun Eom designed, analyzed, and fabricated the sample. Yunsik Seo performed the measurements. Sungjoon Lim conceived the idea and contributed to revising of the manuscript.

CONFLICTS OF INTEREST

The authors declare no conflict of interest.

REFERENCES

1. Qin, P.Y.; Guo, Y.J.; Weily, A.R.; Liang, C.H. A pattern reconfigurable U-slot antenna and its applications in MIMO systems. *IEEE Trans. Antenn. Propag.* **2012**, *6*, 516–528.
2. Lai, M.I.; Wu, T.Y.; Hsieh, J.C.; Wang, C.H.; Jeng, S.K. Design of reconfigurable antennas based on an L-shaped slot and PIN diodes for compact wireless devices. *IET. Micro. Antenn. Propag.* **2009**, *3*, 47–54.
3. Baik, J.W.; Pyo, S.; Lee, T.H.; Kim, Y.S. Switchable printed Yagi-Uda antenna with pattern reconfiguration. *ETRI J.* **2009**, *31*, 318–320. [CrossRef]
4. Lim, J.H.; Back, G.T.; Ko, Y.I.; Song, C.W.; Yun, T.Y. A reconfigurable PIFA using a switchable PIN-diode and a fine-tuning varactor for USPCS/WCDMA/m-WiMAX/WLAN. *IEEE Trans. Antenn. Propag.* **2010**, *58*, 2404–2411.

5. Jung, C.W.; Lee, M.J.; Li, G.P.; De Flaviis, F. Reconfigurable scan-beam single-arm spiral antenna integrated with RF-MEMS switches. *IEEE Trans. Antenn. Propag.* **2006**, *54*, 455–463.
6. Huff, G.H.; Bernhard, J.T. Integration of packaged RF MEMS switches with radiation pattern reconfigurable square spiral microstrip antennas. *IEEE Trans. Antenn. Propag.* **2006**, *54*, 464–469.
7. Kawase, T.; Moriya, S.; Newsome, C.J.; Shimoda, T. Inkjet printing of polymeric field-effect transistors and its applications. *Jpn. J. Appl. Phys.* **2005**, *44*, 3649.
8. Cook, B.S.; Tehrani, B.; Cooper, J.R.; Tentzeris, M.M. Multilayer inkjet printing of millimeter-wave proximity-fed patch arrays on flexible substrates. *IEEE. Antenn. Wirel. Propag. Lett.* **2013**, *12*, 1351–1354.
9. Ahmed, S.; Tahir, F.; Shamim, A.; Cheema, H.M. A compact Kapton-based inkjet printed multiband antenna for flexible wireless devices. *IEEE. Antenn. Wirel. Propag. Lett.* **2015**, *14*, 1802–1805.
10. Hettak, K.; Ross, T.N.; James, R.; Momciu, A.; Wight, J.S. Flexible plastic substrate-based inkjet printed CPW resonators for 60 GHz ISM applications. In Proceedings of the European Microwave Conference (EuMC), Rome, Italy, 6–9 October 2014; pp. 1194–1197.
11. Yang, L.; Rida, A.; Vyas, R.; Tentzeris, M.M. RFID tag and RF structures on a paper substrate using inkjet-printing technology. *IEEE Trans. Microw. Theory Technol.* **2007**, *55*, 2894–2901.
12. Shaker, G.; Safavi-Naeini, S.; Sangary, N.; Tentzeris, M.M. Inkjet printing of ultrawideband (UWB) antennas on paper-based substrates. *IEEE. Antenn. Wirel. Propag. Lett.* **2011**, *10*, 111–114.
13. Yang, L.; Zhang, R.; Staiculescu, D.; Wong, C.P.; Tentzeris, M.M. A novel conformal RFID-enabled module utilizing inkjet-printed antennas and carbon nanotubes for gas-detection applications. *IEEE. Antenn. Wirel. Propag. Lett.* **2009**, *8*, 653–656.
14. McKerricher, G.; Perez, J.G.; Shamim, A. Fully inkjet printed RF inductors and capacitors using polymer dielectric and silver conductive ink with through vias. *IEEE Trans. Electron Dev.* **2015**, *62*, 1002–1009.
15. Zabri, S.N.; Cahill, R.; Conway, G.; Schuchinsky, A. Inkjet printing of resistively loaded FSS for microwave absorbers. *Electron. Lett.* **2015**, *51*, 999–1001.
16. Choi, S.; Su, W.; Tentzeris, M.M.; Lim, S. A novel fluid-reconfigurable advanced and delayed phase line using inkjet-printed microfluidic composite right/left-handed transmission line. *IEEE Microw. Wirel. Compon. Lett.* **2015**, *25*, 142–144.
17. Kim, H.K.; Ling, K.; Kim, K.; Lim, S. Flexible inkjet-printed metamaterial absorber for coating a cylindrical object. *Opt. Express* **2015**, *23*, 5898–5906.

18. Walther, M.; Ortner, A.; Meier, H.; Löffelmann, U.; Smith, P.J.; Korvink, J.G. Terahertz metamaterials fabricated by inkjet printing. *Appl. Phys. Lett.* **2009**, *95*, 251107.
19. Cook, B.S.; Shamim, A. Utilizing wideband AMC structures for high-gain inkjet-printed antennas on lossy paper substrate. *IEEE. Antenn. Wirel. Propag. Lett.* **2013**, *12*, 76–79.
20. Durgun, A.C.; Balanis, C.; Birtcher, C.R.; Allee, D.R. Design, simulation, fabrication and testing of flexible bow-tie antennas. *IEEE Trans. Antenn. Propag.* **2011**, *59*, 4425–4435.
21. George, J.; Deepukumar, M.; Aanandan, C.K.; Mohanan, P.; Nair, K.G. New compact microstrip antenna. *Electron. Lett.* **1996**, *32*, 508–509. [CrossRef]
22. Blech, M.D.; Eibert, T.F. A dipole excited ultrawideband dielectric rod antenna with reflector. *IEEE Antenn. Propag, Trans.* **2007**, *55*, 1948–1954.
23. Sinton, S.; Rahmat-Samii, Y. Random surface error effects on offset cylindrical reflector antennas. *IEEE Trans. Antenn. Propag.* **2003**, *51*, 1331–1337.
24. Garibello, B.E.; Barbin, S.E. A single element compact printed bowtie antenna enlarged bandwidth. In Proceedings of the 2005 SBMO/IEEE MTT-S International Conference on Microwave and Optoelectronics, Brasilia, Brazil, 25–28 July 2005; pp. 354–358.
25. Kim, S.G.; Chang, K. Ultrawide-band transitions and new microwave components using double-sided parallel-strip lines. *IEEE Trans. Microw. Theory Techn.* **2004**, *52*, 2148–2152.
26. Wheeler, H. Transmission-line properties of parallel strips separated by a dielectric sheet. *IEEE Trans. Microw. Theory Techn.* **1965**, *13*, 172–185.
27. Park, J.A.; Cho, C.S.; Lee, J.W. A new design approach for asymmetric coupled-section Marchand balun. *J. Electromagn. Eng. Sci.* **2014**, *14*, 54–60.

5

CHAPTER

QUALITY FACTOR EFFECT ON THE WIRELESS RANGE OF MICROSTRIP PATCH ANTENNA STRAIN SENSORS

Ali Daliri [1,*], Amir Galehdar [1], Wayne S. T. Rowe [1], Sabu John [2], Chun H. Wang [2] and Kamran Ghorbani [1]

[1] *School of Electrical and Computer Engineering, RMIT University, GPO Box 2476, Melbourne, Victoria 3001, Australia*

[2] *School of Aerospace, Mechanical and Manufacturing Engineering, RMIT University, PO Box 71, Bundoora, Victoria 3083, Australia*

ABSTRACT

Recently introduced passive wireless strain sensors based on microstrip patch antennas have shown great potential for reliable health and usage monitoring in aerospace and civil industries. However, the wireless interrogation range of these sensors is limited to few centimeters, which restricts their practical application. This paper presents an investigation on the effect of circular microstrip patch antenna (CMPA) design on the quality factor and the maximum practical wireless reading range of the sensor. The results reveal that by using appropriate substrate materials the interrogation distance of the CMPA sensor can be increased four-fold, from the previously reported 5 to 20 cm, thus improving considerably the viability of this type of wireless sensors for strain measurement and damage detection.

Keywords: wireless strain sensor; microstrip patch antenna; quality factor; structural health monitoring; passive sensor

1. INTRODUCTION

In the last few decades, technological advances and state-of-the-art engineered materials have enabled researchers and engineers to develop several novel methods for Structural Health Monitoring (SHM). These methods are based on a variety of physical phenomena such as eddy current [1], thermal/infrared [2,3] and electromagnetics [4,5]. In addition, sensor based health and usage monitoring systems utilize several types of sensors to monitor strain or crack growth in the structure. These sensors include resistive strain gauges, piezoelectric transducers, fiber optic sensors, and many more. However, the required wiring to connect and power these sensors is a major issue for the broad adoption of SHM. Wireless strain sensors, powered by an external power source/integrated battery unit, have been extensively studied in the literature. The complexity, large size, added weight, and the limited lifetime of batteries (power source) restrict the integration of such sensors in SHM systems.

More recently, the concept of passive wireless strain sensors based on various types of electromagnetic resonators has been introduced to reduce the complexity of traditional wireless sensors by eliminating the need for integrated communication components and power sources [6–16]. Details on the different methods used for developing passive wireless sensors for SHM can be found in two recent review articles [15,16]. Such electromagnetic structures are promising candidates for wireless SHM as their resonant frequency, which is sensitive to their physical dimensions, can be exploited for measurement. These structures include microstrip patch antennas [6–10], Radio Frequency Identification (RFID) tags [11,12], and metamaterial inspired resonators [13,14].

Circular microstrip patch antennas (CMPAs) are one of the promising devices to measure strain in different aerospace materials (both conductive and non-conductive); a linear relationship exists between the strain and the shift in their resonant frequency [8–10]. Recently, a technique to monitor the resonant frequency of CMPAs wirelessly using a linearly polarized horn antenna was demonstrated numerically and experimentally [17]. It was shown that using this technique strain could be monitored in any desired direction by rotating the horn antenna because the CMPA was excited in the direction aligned to the polarization plane of the horn antenna [17]. An important parameter in the wireless measurement of strain using CMPAs is the interrogation distance between the reader antenna and the CMPA sensor. It was reported that by increasing the interrogation distance, the reliability of measuring the shift in the resonant frequency of the CMPA decreases; the maximum practical interrogation distance was found to be 5 cm [17].

In the case of metamaterial resonators [14] and inductive coupling of LC circuits [18] the interrogation distance is limited to a few centimetres. For RFID-based sensors the maximum interrogation distance that has been reported was 1.270 m where curve fitting was required to determine the resonant frequency [12]. For these sensors an additional Integrated Circuit (IC) chip is required which increases the size and complexity of the sensor unit. The maximum interrogation distance for microstrip patch antennas reported in the literature is 1 m where an additional light activated RF switch is required to separate the sensor response from the structural response [6]. In a recent study using dipole antennas as the sensing element the maximum interrogation distance reported is 15 cm [19]. However, the dipole element has been attached to a dielectric where the reflections from the structure are not significant. Until now, no detailed study of the effect of antenna design parameters on the interrogation distance of passive antenna strain sensors, without using additional circuit elements, has been reported.

The wireless reading range of CMPAs needs to be increased to enable the practical implementation of this technique for SHM. In this paper, we present a study of the effect of CMPA's quality factor (which is a representation of the losses in the CMPA) on the interrogation distance. First, the quality factor of microstrip patch antennas and the effect of various antenna parameters on the quality factor are studied. Based on this study, a CMPA with high quality factor is designed and the effect of antenna quality factor on the reading range is discussed. Finally, the CMPA with high quality factor is fabricated and its improved performance in wireless strain measurement is validated experimentally. The results show that by using high quality factor antennas/resonators the wireless reading range can be increased significantly.

2. THEORETICAL CONSIDERATIONS AND COMPUTATIONAL ANALYSES

The fractional bandwidth of an antenna is inversely proportional to the quality factor of the antenna and it can be expressed as [20]:

$$\text{Fractional Bandwidth} = \frac{\Delta f}{f} = \frac{1}{Q_t} \quad (1)$$

A more precise relationship can be defined by accounting for the impedance matching of the antenna [20]. However, for the wireless excitation of a microstrip patch antenna this impedance matching dependence is not required because the patch antenna is not excited using a transmission line (such as a coaxial cable). The quality factor of an antenna can be defined as a representation of the antenna losses [20]. According to [21] the total quality factor for a circular microstrip patch antenna can be calculated from the following equation:

$$\frac{1}{Q_t} = \frac{1}{Q_{rad}} + \frac{1}{Q_c} + \frac{1}{Q_d} + \frac{1}{Q_{sw}} \tag{2}$$

Figure 1 shows the geometry of the CMPA sensor. For very thin substrates ($h << \lambda_0$), the loss due to surface waves, $1/Q_{sw}$, is very small and can be neglected in calculation of the total quality factor [21]. This indicates that by reducing the thickness of the patch antenna the quality factor due to the surface waves can be improved and as a result of that the total quality factor of the antenna can be improved. The other quality factors can be calculated using following equations [21]:

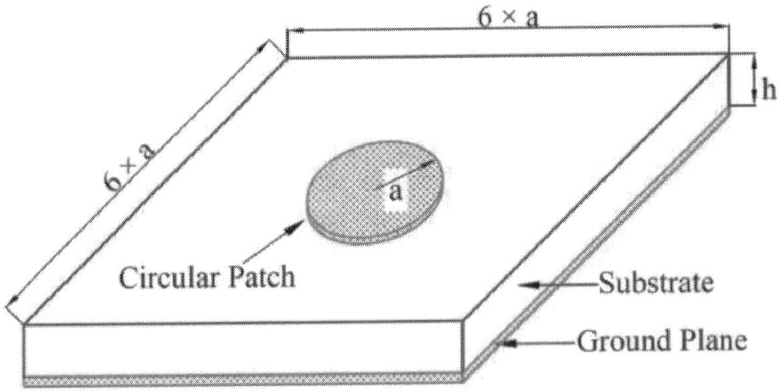

Figure 1. Geometry of the CMPA sensor.

$$Q_c = h\sqrt{\pi f \mu_0 \sigma} \tag{3}$$

$$Q_d = \frac{1}{\tan\delta} \tag{4}$$

and for the dominant mode of operation [22]:

$$Q_{rad} = \frac{30[(ka)^2 - 1]}{hf\mu_0 (k_0 a)^2 I_1} \tag{5}$$

where:

$$I_1 = \int_0^{\pi/2} [J_1'^2(k_0 a \sin\theta) + \cos^2\theta J_1^2(k_0 a \sin\theta)/(k_0 a \sin\theta)^2] \sin\theta d\theta \tag{6}$$

The quality factor due to dielectric losses can be improved by using low loss materials for the antenna substrate. From Equation (3), the conductive loss can be reduced by increasing the conductivity of the patch and the ground plane. In practice, the Q_c will be lower than Equation (3) because of the surface roughness of the patch and the ground plane. According to [20], for very thin substrates the dominant factor is the radiation quality factor. This part of the total quality factor is proportional to the substrate dielectric constant and the inverse of the substrate thickness.

2.1. Effect of Substrate Material on Quality Factor

In order to improve the quality factor of the CMPA, a numerical investigation into the effect of each parameter of the total quality factor was carried out. To this end, two commercially available substrates were selected, namely a FR4 (ε_r = 4.5, tanδ = 0.025) and Rogers™ RT/duroid 6010.2LM™ (ε_r = 10.2, tanδ = 0.0023 [23]; Rogers Corporation, Brooklyn, CT, USA). The FR4 substrate was selected because it was used in previous studies on CMPA sensors [8–10,17] and the Rogers™ substrate (which is called high Q material in the rest of the paper) was used because it has much lower tangent loss and much higher permittivity compared to FR4. For the numerical study the thickness of each substrate was varied within the range of commercially available laminate thicknesses, i.e., 0.127, 0.254, 0.635, 1.27, 1.90, 2.50 mm for high Q material and 0.8, 1.0, 1.2, 1.5, 2.0, 2.4 mm for FR4. The antennas were designed to resonate at 1.5 GHz (similar to previous studies on CMPAs [8–10,17]).

The conductor, dielectric and radiation quality factors for the two substrates were calculated using Equations (3)–(6) and the results for different substrate thicknesses are shown in Figure 2. The results show that the conductor quality factor is independent of the substrate permittivity and varies linearly with the substrate thickness. From Equation (4), the dielectric quality factor depends only on the loss tangent of the substrate.

The results in Figure 2 shows for high Q material the dielectric quality factor is much higher than the FR4 substrate. The radiation quality factor depends on both the substrate thickness and the substrate material's permittivity. The results in Figure 2 show that radiation quality factor is higher for the substrate with the higher dielectric constant regardless of the substrate thickness. By reducing the substrate thickness of the high Q material the radiation quality factor increases dramatically.

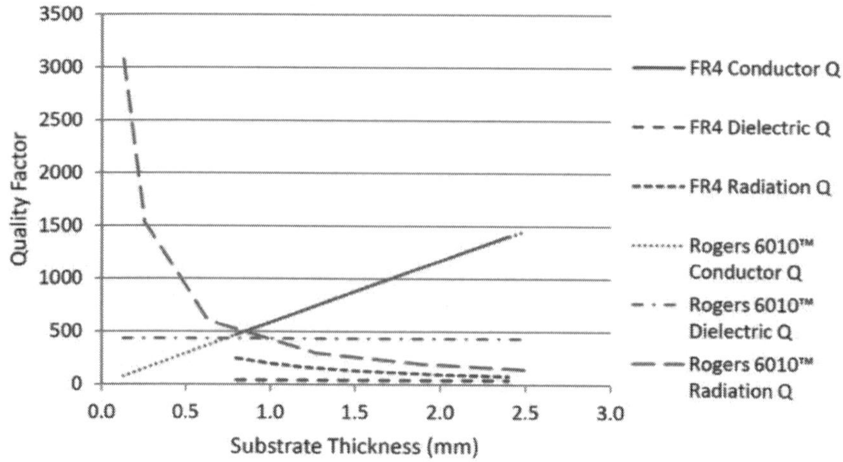

Figure 2. Conductor, dielectric and radiation quality factors for two substrate materials studied with different thicknesses.

Figure 2 indicates that the substrate with higher permittivity and lower loss tangent will always have a higher total quality factor regardless of the substrate thickness. This can also be seen from Figure 3 where the total quality factors for the two substrates, according to Equation (2), are compared with the numerical results from HFSS™ (Eigen mode solver) (ANSYS, Canonsburg, PA, USA) [24]. Substrate thickness has opposite effects on the radiation and conductor quality factors. Figure 3 shows that for commercially available thicknesses of high Q material, the optimum thickness to achieve the highest total quality factor is 0.635 mm. Therefore, this thickness was chosen for further simulations and fabrication of the CMPA sensor to be employed in experimental measurements.

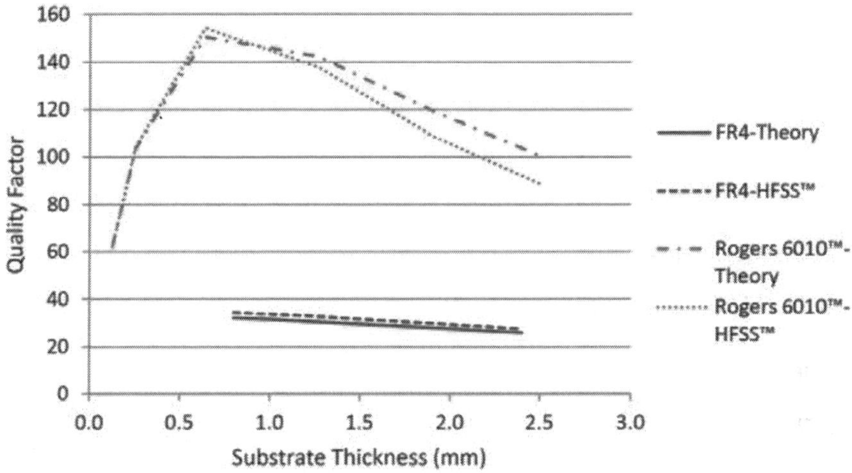

Figure 3. Total quality factor for the two substrates studied at different thicknesses.

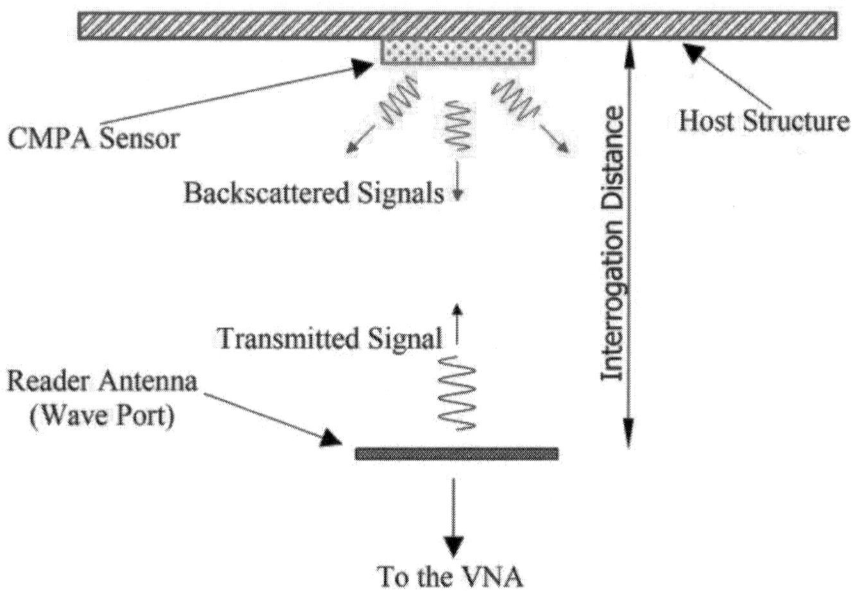

Figure 4. Schematic of the wireless setup for simulation and measurement.

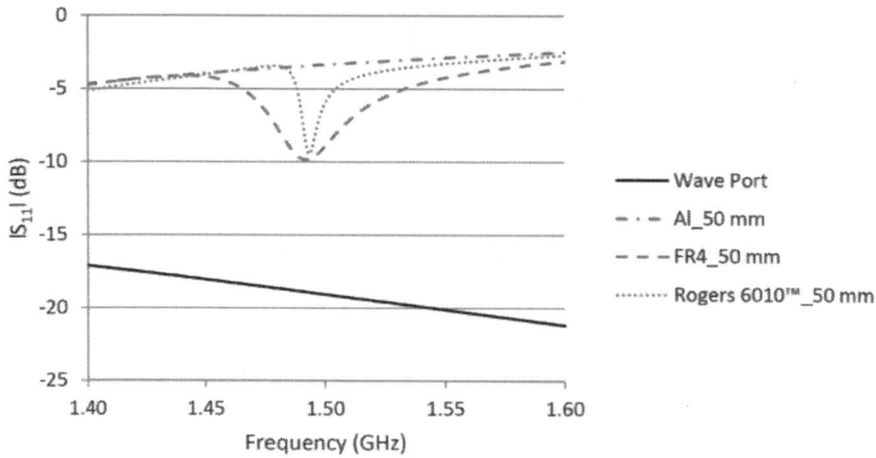

Figure 5. Effect of quality factor on the $|S_{11}|$ of two CMPAs.

2.2. Effect of Quality Factor on Interrogation Distance

To investigate the effect of the quality factor on the interrogation distance of a CMPA strain sensor, two CMPAs were analyzed in HFSS™ with a setup similar to the simulation schematic in [17]. To reduce the computational time, the horn antenna used in [17] was replaced with a rectangular wave port with the same aperture where only mode 1 of the wave port was excited. Figure 4 shows the diagram of the setup used in the simulations and measurements. Figure 5 presents the results of $|S_{11}|$ of the wave port for the interrogation distance of 5 cm when CMPA was attached to an aluminum plate and placed in front of the wave port. In addition, the $|S_{11}|$ of the wave port without an object and with an aluminum plate in front of it are shown.

As discussed in [17], the reflection from the aluminum plate increases the $|S_{11}|$ at the wave port because some of the incident energy is reflected back to the port. The magnitude of the $|S_{11}|$ depends on the distance between the aluminum plate and the wave port resulting from a standing wave. Regardless of the $|S_{11}|$ magnitude for the aluminum plate, by attaching a CMPA to the aluminum plate the resonant frequency of the CMPA can be clearly distinguished. By increasing the interrogation distance the difference between the $|S_{11}|$ magnitude from the bare aluminum plate and the $|S_{11}|$ at the resonant frequency of the CMPA become less, which makes it difficult to detect the resonant frequency of the CMPA. Figure 5 shows that for the same interrogation distance and same resonant frequency, the dip of

$|S_{11}|$ is much sharper (meaning the CMPA bandwidth is smaller) for the high Q substrate compared to the FR4 design. This is the effect of the higher quality factor of the CMPA designed using the high Q material. The sharper response of the high Q CMPA results in easier and more accurate tracking of the shift of the CMPA resonant frequency due to strain, because the sharper response has one clear minimum.

Figure 6 shows the effect of interrogation distance on the $|S_{11}|$ of both CMPAs attached to the aluminum plate. By changing the interrogation distance the standing wave between the aluminum plate and the wave port changes and accordingly affects the magnitude of the $|S_{11}|$. However, regardless of the level of $|S_{11}|$ for the aluminum plate, the resonant frequency of the CMPA generates a dip in the $|S_{11}|$ curve. The relative magnitude of this dip decreases with the interrogation distance for both CMPAs. Beyond a certain interrogation distance this relative dip vanishes, making it very difficult to detect the resonant frequency of the CMPA. Figure 6 shows that for the CMPA designed with high Q material because of a sharper dip in the $|S_{11}|$ curve the resonant frequency of the CMPA can be distinguished from the structure response up to 20 cm whilst for the CMPA designed with FR4 it is more difficult to distinguish the resonant frequency even at 10 cm.

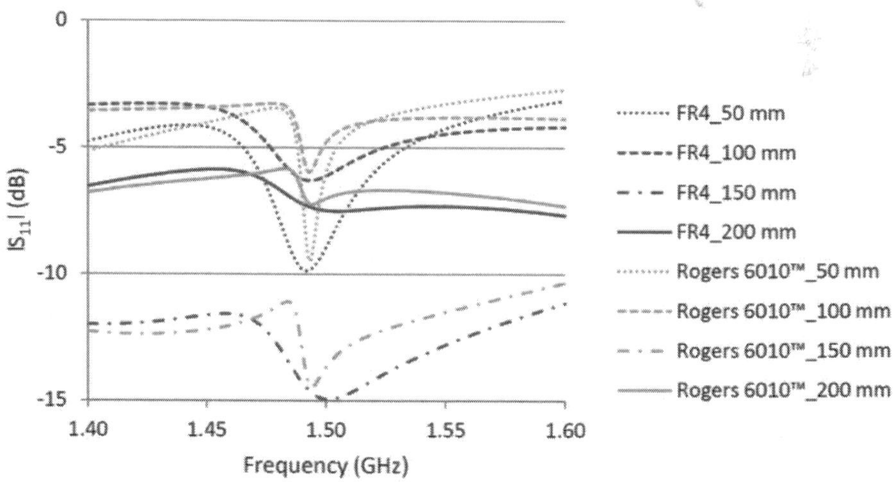

Figure 6. Effect of interrogation distance on the $|S_{11}|$ of two CMPAs.

3. EXPERIMENTAL MEASUREMENTS

To validate the feasibility of increasing wireless reading range using high quality factor CMPAs, the proposed high Q CMPA was fabricated and its strain response was measured. The experimental setup was similar to that employed in [17]. A linearly polarized horn antenna (DRH118™ horn antenna from Sunol Sciences® [25]) (Sunol Sciences Corporation, Dublin, CA, USA) was used as the reader antenna. In order to apply strain to the host structure, a tensile force of 20 kN was applied to the assembly in 20 load increments (1 kN steps) using a 100 kN INSTRON™ machine (Instron, Norwood, MA, USA). At the maximum applied load, the strain in the center of the host structure is 500 με. Nine conventional resistive strain gauges were attached to the back of the host structure to verify that a uniform strain is applied in the region where the CMPA is attached. The $|S_{11}|$ of the horn antenna and the strain were measured at each loading step. The resonant frequency of the CMPA has been extracted by reading the minima in the $|S_{11}|$ curve as well as the smith chart (real and imaginary parts). The network analyzer was set to measure 1601 points in the measurement frequency range, resulting in a resolution of 25 με per 25 kHz increment. Due to slight differences in the resonant frequencies, the measurement frequency range for the FR4 and high Q CMPAs were 1.535 GHz to 1.575 GHz, and 1.470 GHz to 1.510 GHz, respectively. Reducing the frequency increments below 25 kHz was found to produce no improvement in the sensitivity of the sensor.

For both CMPAs, the tests were conducted for interrogation distances of 1, 5, 10, 15 and 20 cm. The wireless measurement of strain for the CMPA made from FR4 was only possible at interrogation distances of up to 5 cm. By increasing the interrogation distance the $|S_{11}|$ curve de-stabilizes and the minima moves back and forth. These variations in the $|S_{11}|$ were greater than the shift in the resonant frequency (which is caused by the applied strain). This resulted in a decrease in the signal to noise ratio of the wireless strain measurement. The relationships between the strain and the shift in the resonant frequency of the CMPA with FR4 substrate at two different interrogation distances are shown in Figure 7. A linear relationship between the applied strain and the shift in the resonant frequency is observed for 1 and 5 cm interrogation distances. At interrogation distances beyond 5 cm wireless strain measurement was no longer possible for the CMPA made from FR4, because of the increased variations in the $|S_{11}|$.

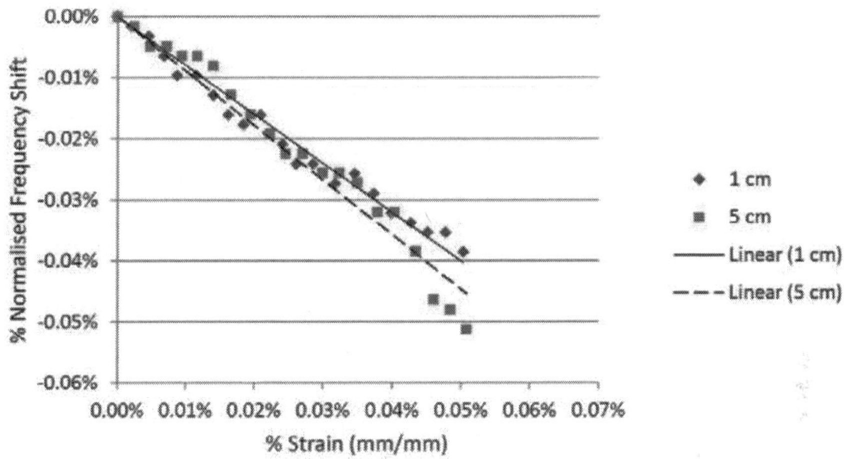

Figure 7. Strain-frequency shift relationship for the FR4 CMPA sensor.

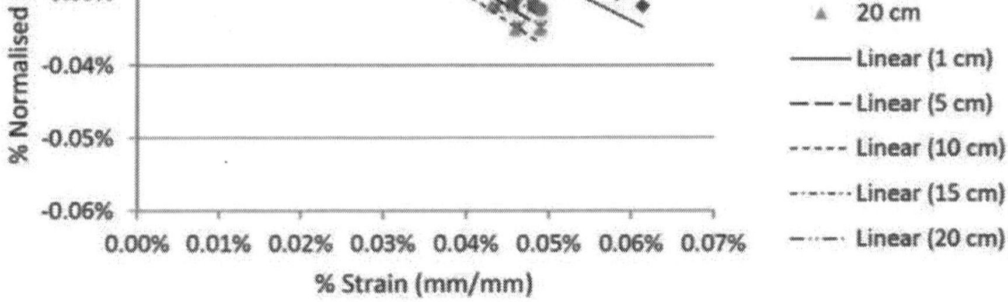

Figure 8. Strain-frequency shift relationship for the Rogers 6010™ CMPA sensor.

For the CMPA made from high Q material the wireless strain measurement remained reliable at interrogation distances of up to 20 cm, albeit with a slight decrease in resolution at longer distances. Figure 8 shows the relationship between the strain and the shift in the resonant frequency of the CMPA with the high Q material at interrogation distances of 1, 5, 10, 15 and 20 cm. The results for the 20 cm interrogation distance show the increased wireless reading range using this substrate, confirming the findings from the computational simulation presented in the previous section. Therefore, utilizing a substrate with higher permittivity and lower loss tangent with the right thickness improves the wireless range of the CMPA because of increased quality factor. It is envisaged that increasing gain and efficiency of the antenna sensor and the reader antenna will also improve the interrogation distance.

4. CONCLUSIONS

The effect of quality factor on the interrogation distance of CMPA strain sensors was investigated using numerical calculations and experimental measurements. The results presented in this paper show that the quality factor of the CMPA has a significant effect on the wireless reading range. Strain measurement at interrogation distances up to 20 cm was possible using the high Q material where previous studies using FR4 CMPA were only able to measure strain at 5 cm. Therefore the interrogation distance can be increased by a factor of 4 when a substrate with lower loss and higher permittivity is employed. The analysis presented in this paper was based on commercially available substrates. Further increase in the interrogation distance is possible by using specially designed materials that exhibit significantly low losses and very high dielectric constants.

CONFLICTS OF INTEREST

The authors declare no conflict of interest.

REFERENCES

1. García-Martín, J.; Gómez-Gil, J.; Vázquez-Sánchez, E. Non-destructive techniques based on eddy current testing. Sensors **2011**, 11, 2525–2565.

2. Vavilov, V. Noise-limited thermal/infrared nondestructive testing. NDT&E Int. **2014**, 61, 16–23.

3. Schäfer, T.; Schubert, M.; Hampel, U. Temperature grid sensor for the measurement of spatial temperature distributions at object surfaces. Sensors **2013**, 13, 1593–1602.

4. Mukhopadhyay, S.C. Novel planar electromagnetic sensors: Modeling and performance evaluation. Sensors **2005**, 5, 546–579.

5. Deng, Y.; Liu, X. Electromagnetic imaging methods for nondestructive evaluation applications. Sensors **2011**, 11, 11774–11808.

6. Deshmukh, S.; Huang, H. Wireless interrogation of passive antenna sensors. Meas. Sci. Technol. **2010**, 21.

7. Mohammad, I.; Gowda, V.; Zhai, H.; Huang, H. Detecting crack orientation using patch antenna sensors. Meas. Sci. Technol. **2012**, 23.

8. Daliri, A.; John, S.; Galehdar, A.; Rowe, W.S.T.; Ghorbani, K. Strain Measurement in Composite Materials Using Microstrip Patch Antennas. Proceedings of the ASME Conference on Smart Materials, Adaptive Structures and Intelligent Systems, Philadelphia, PA, USA, 28 September–1 October 2010; Brei, D., Ed.; American Society of Mechanical Engineers: New York, NY, USA, 2010; pp. 1–8.

9. Daliri, A.; Galehdar, A.; Rowe, W.S.T.; Ghorbani, K.; John, S. Utilising microstrip patch antenna strain sensors for structural health monitoring. J. Intell. Mater. Syst. Struct. **2012**, 23, 169–182.

10. Daliri, A.; Galehdar, A.; John, S.; Rowe, W.S.T.; Ghorbani, K. Slotted Circular Microstrip Patch Antenna Application in Strain Based Structural Health Monitoring. Proceedings of the 7th DSTO International Conference on Health and Usage Monitoring (HUMS), Melbourne, Australia, 28 February–3 March 2011; Forrester, D., Forsyth, G., Eds.; Defence Science and Technology Organisation: Melbourne, Australia, 2011; pp. 1–12.

11. Occhiuzzi, C.; Paggi, C.; Marrocco, G. Passive RFID strain-sensor based on meander-line antennas. IEEE Trans. Antennas Propag. **2011**, 59, 4836–4840.

12. Yi, X.; Wu, T.; Wang, Y.; Leon, R.T.; Tentzeris, M.M.; Lantz, G. Passive wireless smart-skin sensor using RFID-based folded patch antennas. Int. J. Smart Nano Mater. **2011**, 2, 22–38.

13. Ekmekci, E.; Turhan-Sayan, G. Metamaterial Sensor Applications Based on Broadside-Coupled SRR and V-Shaped Resonator Structures. Proceedings of the IEEE International Symposium on Antennas and Propagation (APSURSI), Spokane, WA, USA, 3–8 July 2011; pp. 1170–1172.

14. Melik, R.; Unal, E.; Perkgoz, N.K.; Puttlitz, C.; Demir, H.V. Metamaterial-Based Wireless RF-MEMS Strain Sensors. Proceedings of the IEEE Sensors Conference, Kona, HI, USA, 1–4 November 2010; pp. 2173–2176.

15. Deivasigamani, A.; Daliri, A.; Wang, C.H.; John, S. A review of passive wireless sensors for structural health monitoring. Mod. Appl. Sci. **2013**, 7, 57–76.

16. Huang, H. Flexile wireless antenna sensor: A review. IEEE Sensors **2013**, 13, 3865–3872.

17. Daliri, A.; Galehdar, A.; John, S.; Wang, C.H.; Rowe, W.S.T.; Ghorbani, K. Wireless strain measurement using circular microstrip patch antennas. Sens. Actuators A Phys. **2012**, 184, 86–92.

18. Zhang, C.; Huang, J.Q.; Huang, Q.A. A Passive Wireless Graphene Oxide Based Humidity Sensor and Associated Portable Telemetry Unit. Proceedings of the 17th International Conference on Solid-State Sensors, Actuators and Microsystems (Transducers & Eurosensors XXVII), Barcelona, Spain, 16–20 June 2013; pp. 278–281.

19. Sang-Dong, J.; Kim, J. Passive wireless structural health monitoring sensor made with a flexible planar dipole antenna. Smart Mater. Struct. **2012**, 21, 027001.

20. Balanis, C.A. Antenna Theory Analysis and Design, 3rd ed.; John Wiley & Sons: Hoboken, NJ, USA, 2005.

21. Carver, K.R.; Mink, J.W. Microstrip antenna technology. IEEE Trans. Antennas Propag. **1981**, 29, 2–24.

22. Garg, R.; Bhartia, P.; Bahl, I.; Ittipiboon, A. Microstrip Antenna Design Handbook; Artech House: Boston, MA, USA, 2001.

23. Rogers Corporation Webpage. Rogers® RT/duroid 6010™ Datasheet. Available online: http://www.rogerscorp.com/documents/612/acm/RT-duroid-6006-6010-laminate-data-sheet.aspx (accessed on 27 December 2013).

24. ANSYS® HFSS™ HPC. ANSYS 15.0 Release; ANSYS: Canonsburg, PA, USA, 2013.

25. Sunol Sciences® Corporation Webpage. DRH118™ Horn Antenna Datasheet. Available online: http://www.sunolsciences.com/pdf/DRH%20Antennas.pdf (accessed on 27 December 2013.

6

CHAPTER

A COMPACT ANNULAR RING MICROSTRIP ANTENNA FOR WSN APPLICATIONS

Daihua Wang [1,*], Linli Song [2], Hanchang Zhou [2] and Zhijie Zhang [1]

[1] *National Key Laboratory for Electronic Measurement Technology, North University of China, Taiyuan 030051, China*

[2] *Key Laboratory of Instrumentation Science and Dynamic Measurement, Ministry of Education, North University of China, Taiyuan 030051, China*

ABSTRACT

A compact annular ring microstrip antenna was proposed for a wireless sensor network (WSN) application in the 2.4 GHz band. In this paper the major considerations of the conformal antenna design were the compact size and the impact on antenna's performance of a steel installation base. By using a chip resistor of large resistance (120 Ω) the antenna size was reduced to 38% of that a conventional annular ring patch antenna. With the addition of the steel installation base the resonant frequency of the antenna increases about 4.2% and the bandwidth reduces from 17.5% to 11.7% by adjusting the load resistance simultaneously. Several key parameters were discussed and optimized, and the antenna was fabricated and its performance measured. The antenna is well matched at 2.4 GHz with 34.2 dB return loss and –2.5 dBi peak gain. Meanwhile, it exhibits excellent radiation patterns with very low cross-polarization levels.

Keywords: microstrip antenna; resistor loading; effects of metal on antenna characteristics

1. INTRODUCTION

Microstrip patch antennas have been widely used in many systems [1–4] because of their attractive features such as compact size, low profile, light weight and ease of fabrication [5–9]. Various shapes of the patches were proposed for these antennas, such as circular, square, triangular, elliptical, diamond, etc. [6–8,10,11]. However, in some applications the small size conventional patch antenna is still too large [5,12–14], so research is still being focused on the miniaturization of the patch antenna over the years. At present, there were many effective size reduction methods, including using shorting pins, resistor load, high dielectric constant materials and slots, etc. [5–8,15–18].

In this paper, conformal antenna design with compact size was one major consideration. The antenna was proposed for a wireless sensor network (WSN) application in the 2.4 GHz band. The sensor nodes executed monitoring in an explosive experiment for military use. The structure for the antenna installation was designed as a part shell of the sensor node which is illustrated in Figure 1. The materials of all components were steel to resist physical damage in the explosive experiment. The wireless transceiver module was installed beneath the installation base of the antenna and protected by a surrounding steel shell. To propagate out RF signals from the cage the only approach was a coaxial cable through a cylindrical tunnel on the installation base. The patch antenna could be installed only in a thin annular slot on the top of the installation base whose inner radius was greater than 5 mm and outer radius less than 20 mm. Therefore, the size of the designed antenna must comply with the requirements of the limited mounting space.

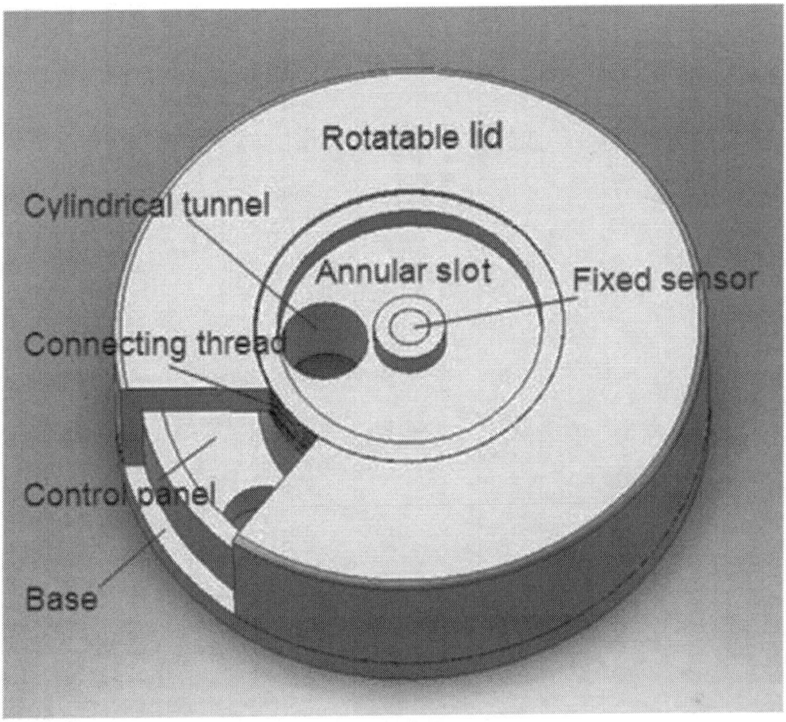

Figure 1. Installation base of the antenna.

Another main consideration of the research was the impact on the antenna's performance of the steel installation base. In recent years the influences of metal carriers on conformal antenna design were studied [19,20], however, there were less concerns about planar antenna design. For most reported microstrip antennas the surroundings were usually considered free space. In our studies the steel installation base greatly influenced the characteristics of the proposed antenna such as resonant frequency, bandwidth and radiation characteristics, etc., so how to design the metal carrier was an important issue in the antenna design.

The configuration of the proposed antenna and the parameter settings are described in Section 2. To meet the compact and conformal requirements an annular ring patch antenna with a loaded chip resistor was designed. In Section 3, key parameters were studied, especially for the outer radius of the substrate and the patch, position of the loading point, and the resistance of the chip resistor. Then the proposed antenna was manufactured and the measured results were compared with the simulation ones. Finally, the conclusions are summarized in Section 4.

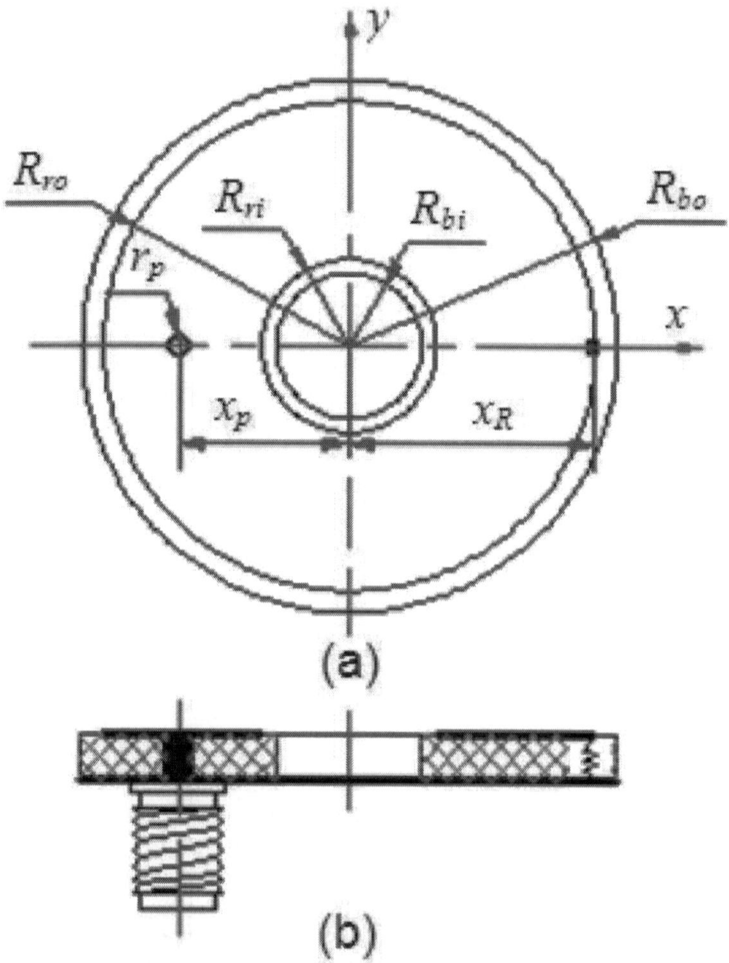

Figure 2. Configuration and parameters of the proposed antenna. (**a**) Top view. (**b**) Cross section.

2. ANTENNA CONFIGURATION AND DESIGN

The schematic of the antenna configuration is shown in Figure 2. It was designed as an annular ring structure with outer radius of R_{bo} (≤20 mm), inner radius of R_{bi} (≥5 mm), and thickness of h. To reduce costs FR4 substrate with relative dielectric constant of $\varepsilon_r = 4.5$ and loss tangent of $\tan\delta = 0.02$ was selected. On the top side of the substrate, an annular ring radiating patch with outer radius of R_{ro} inner radius of R_{ri} and metallization thickness of 35 μm was printed. The ground plane was placed full size on the opposite side of the substrate. As shown in Figure 2 the antenna was back fed by a coaxial probe of radius r_p located at a distance of x_p away from the circular center. The cylindrical tunnel in the installation base shown in Figure 1 was designed as the access for RF signal propagation through the coaxial probe. The diameter of the cylindrical tunnel was just suit for the coaxial probe, while almost as large as the width of the annular slot due to the compact space. So the distance of x_p was fixed around the center of the annular slot. It was shown in [14,21] that for minimum patch area the feed point must be located nearest to the radiating edges of the patch. The criterion in our project cannot be achieved, which brings more burden to the miniaturization.

As a conventional annular ring patch antenna with parameter settings above mentioned, the outer radius of R_{bo} is at least 30 mm at resonant frequency of 2.4 GHz with very narrow bandwidth (as low as 1.7%). It is obvious that a conventional annular ring patch antenna is not suitable for the limited mounting space. As mentioned in Section 1, loading is an effective way to reduce antenna size. In this design, a chip resistor of resistance R was loaded along the x-axis with a distance of x_R away from the circular center and the parameters R and x_R were selected carefully for impedance matching.

The thickness of the substrate (h) was another key parameter for design. Increasing the thickness the bandwidth and the efficiency of the patch antenna was improved [22], while the unwanted surface waves were further excited too [14]. In view of the limited mounting space and the factors above mentioned, the thickness of the proposed antenna was fixed at h = 3 mm.

3. RESULTS AND DISCUSSION

To optimize antenna's characteristics, such as resonant frequency, bandwidth, gain, radiation efficiency and level of cross-polarization, a series of parameters of the annular ring antenna should be carefully selected. In this section, the effects of several key parameters on the proposed antenna were discussed and simulated using the Ansoft HFSS. In comparison with the simulated results, several antennas were fabricated and measured.

3.1. Effects of Key Parameters in Free Space

3.1.1. Effect of the Antenna Dimensions

Figure 3 shows the simulated return loss plots for different dimensions of the annular ring with fixed parameters as R_{bi} = 5 mm, R_{bo} = 18.5 mm, r_P = 0.6 mm, x_p = −12 mm, x_R = R_{ro} − 0.5 mm, and R = 56 Ω. From the simulated results of Figure 3, it is found that the resonant frequency is very sensitive to the patch size. As the outer radius (R_{ro}) or inner radius (R_{ri}) increases, the resonant frequency is decreased. The antenna impedance is well matched at 2.4 GHz with the combination of R_{ri} = 5.5 mm and R_{ro} = 17 mm, and the impedance matching is degraded while the patch dimensions deviate from the optimum values. Compared with the conventional annular ring patch antenna above mentioned, Figure 3 shows that a greater bandwidth can be obtained. There is also a notable decrease in the bandwidth as the outer radius (R_{ro}) increases, while the inner radius (R_{ri}) causes negligible effects.

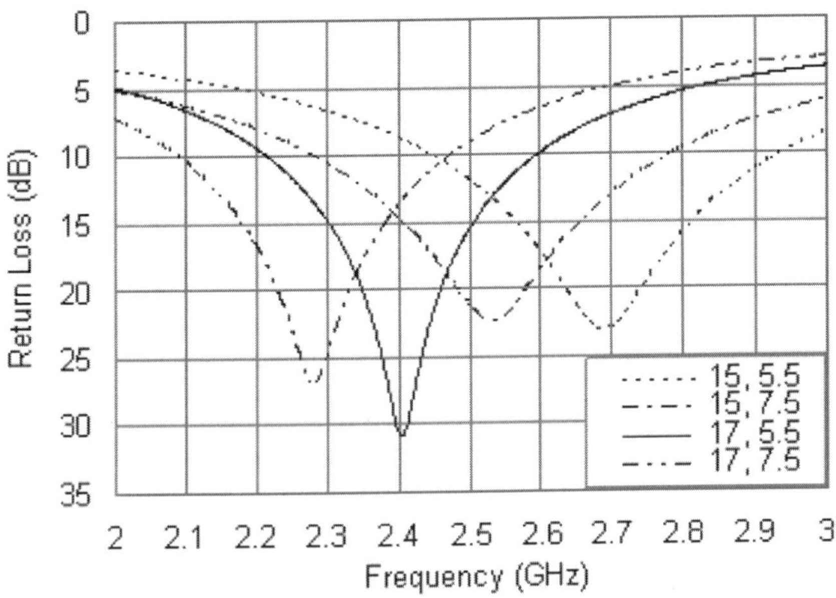

Figure 3. Simulated return loss curves with different values of R_{ro} and R_{ri}.

In the process of analysis, it is also found that in comparison with the patch size the substrate size has relatively less effects on the resonant frequency. The resonant frequency slightly decreases with increasing outer radius (R_{bo}). To meet the compact requirements of the antenna, the outer radius of the substrate (R_{bo}) was selected as 18.5 mm with consideration of the optimized performance. The inner radius of the substrate (R_{bi}) was fixed at 5 mm due to the installation requirement of the sensor (shown in Figure 1).

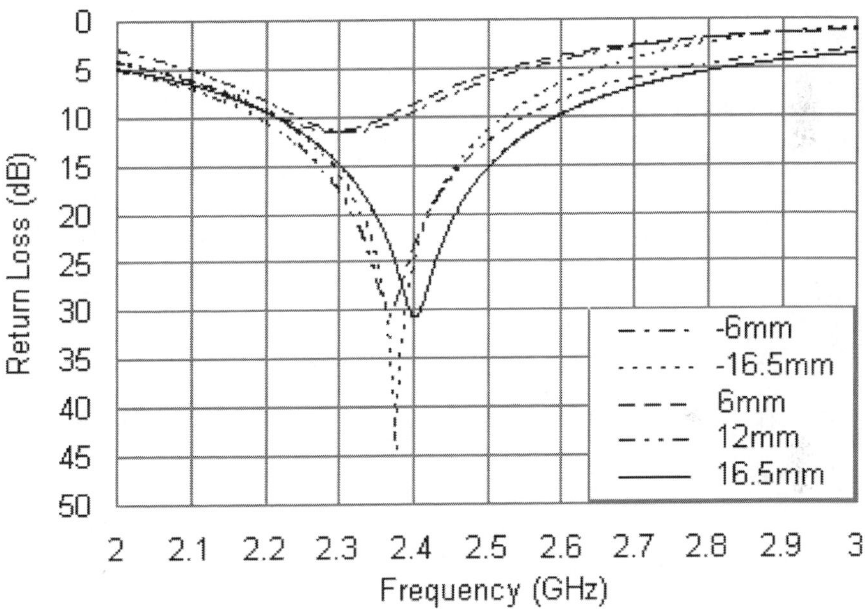

Figure 4. Simulated return loss curves with different values of x_R.

3.1.2. Effect of the Loading Position

To minimize the dimensions of the proposed antenna and increase the operating bandwidth, a chip resistor was introduced into the design to adjust the input impedance characteristics. When modeling in HFSS, a thin two-dimensional surface (0.3 mm × 3 mm) was placed in parallel with the z-axis at the loading position between the patch and the ground, and a lumped RLC boundary was assigned in parallel on the surface. Figure 4 shows the return loss curves for different positions of the loading point with other parameters of R_{ri} = 5.5 mm, R_{ro} = 17 mm, r_P = 0.6 mm, x_p = −12 mm and R = 56 Ω. There is an upward shift for the resonant frequency as the loading point is moved away from the circular center with the bandwidth improved significantly, and a notable agreement between the loading points at same radius along the x-axis is also observed. It was shown in [14,21] that for minimum area the loading point should be located closest to one of the radiating edges of the patch. As seen from Figure 4, the case of loading at the inner edge of the patch can result in a further reduction in resonant frequency, which corresponds to an even larger size reduction at a given operating frequency. In this case, the resistor loading has maximum effects on the resonant-frequency lowering due to the surface current has a maximum value around the inner edge of the patch, however, the matching level is poor. With consideration of the symmetrical characteristics [23] and easy of fabrication, the loading point was placed at the outer edge of the patch with x_R = 16.5 mm.

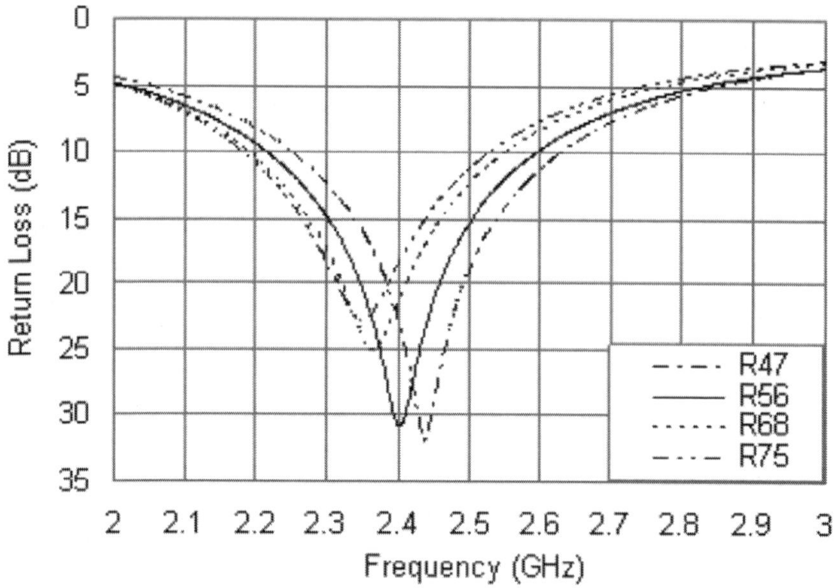

Figure 5. Simulated return loss curves with different values of R.

3.1.3. Effect of the Loading Resistance

Another important parameter is the resistance of the loading resistor. Efforts were made to optimize the value of R to obtain optimum impedance matching for the proposed antenna. Figure 5 shows the return loss curves for different values of R and fixing other parameters as: R_{ri} = 5.5 mm, R_{ro} = 17 mm, r_P = 0.6 mm, x_p = −12 mm and x_R = 16.5 mm. It is evident that the resonant frequency of the antenna is dependent on the resistance of the loading resistor. As the R value increases, the center frequency decreases with the matching level slightly affected. No significant change in bandwidth is observed with the changes of R. Hence, the value of R can be adjusted to provide better resonant mode and good matching.

In comparison with the previously reported designs [24–26], an important change of this design, which kept the antenna operating at required frequency while allowing dramatically reduced dimensions, was the large resistance of the loading resistor. Several resistors with small resistance less than 10 Ω were analyzed, and the resonant frequencies were higher than that one in Figure 5 with poor matching levels. For example, the center frequency is 2.65 GHz as the resistance of R = 1 Ω, and the return loss is as low as 4.7 dB. In this case, the matching conditions are destroyed, and the impedance is mismatched. However, it should be noted that to reduce antenna dimensions the introducing resistor of large resistance was at the expense of the radiation efficiency, similar to the observation in [25–27]. The radiation efficiency of the conventional annular ring antenna above mentioned is 24.9%, with the introducing resistor (R = 56 Ω), the radiation efficiency of the proposed antenna is reduced to 14.6% and the peak gain is −1.5 dBi.

3.2. Measurement and Comparison

The fabricated prototype of the proposed antenna is shown in Figure 6. A chip resistor in 0402 package was mounted directly onto the bottom surface of the PCB. The loading terminal of the chip resistor was connected to the radiating patch through a small via hole, and the opposite side was directly connected to the ground plane. The via hole was drilled at the position of x_R = 16.5 mm and with a diameter was 0.3 mm which complied with the simulation settings. The only variation between the fabricated prototype and the simulation model was the mounting method of the resistor, and the following results showed that the influence of the variation was negligible. The measured and simulated return loss curves are compared in Figure 7. Both the measured and simulated curves are excited at 2.4 GHz with return loss as 56.7 and 30.8 dB, respectively. The measured impedance bandwidth is 17.5% from 2.20 to 2.62 GHz, and the simulated impedance bandwidth is 16.3% from 2.21 to 2.60 GHz. The measured results show good agreement with the simulated ones in Figure 7.

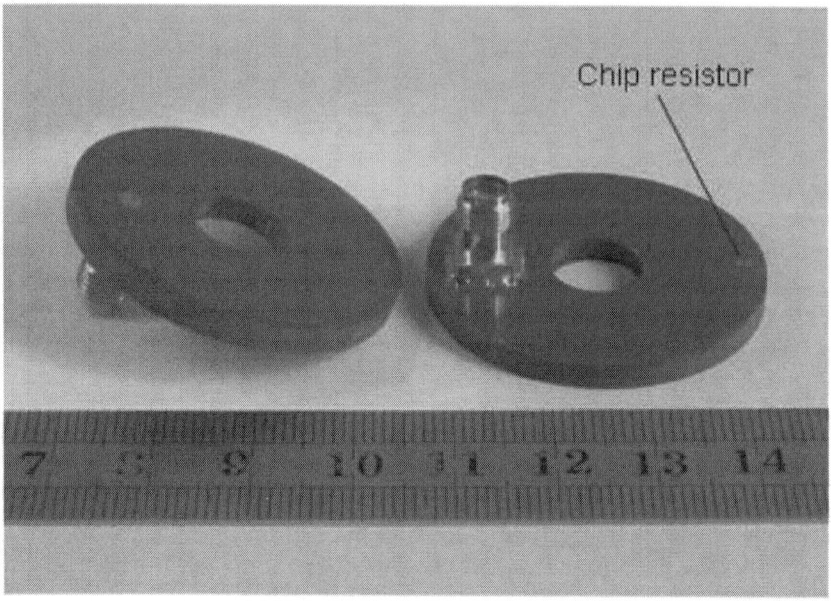

Figure 6. Photograph of the fabricated antenna.

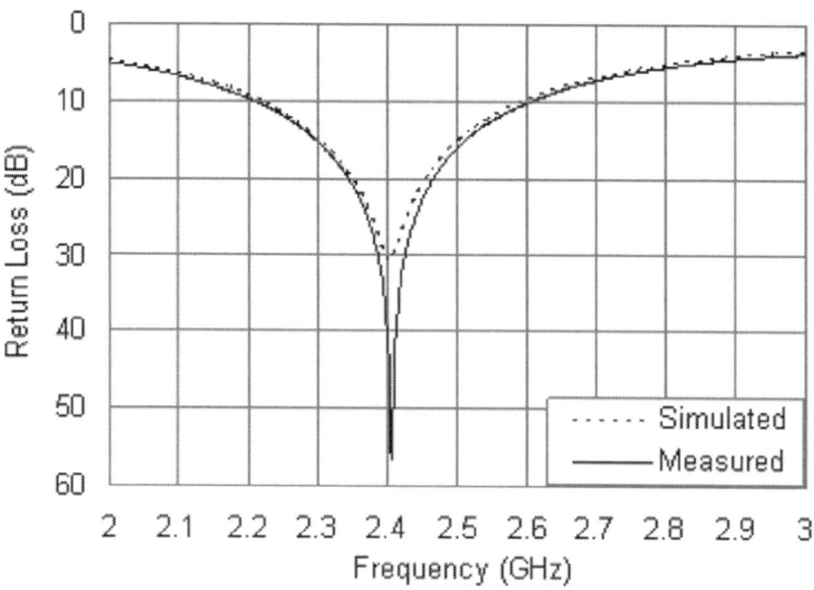

Figure 7. Measured and simulated return loss curves.

Figure 8 presents the measured and simulated radiation patterns including the co-polarization and cross-polarization in the E- and H-plane. From the measured results the radiation patterns are symmetric and there is minimal ripple in both E- and H-plane which are in good agreement with the simulation. The antenna also shows good cross-polarization level as low as −20 dB. However, due to a small ground-plane size, backward radiations are observed obviously in both planes.

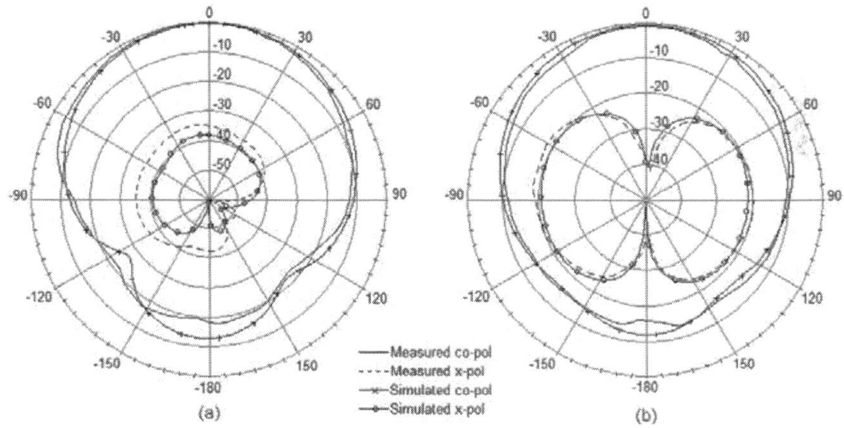

Figure 8. Measured and simulated radiation patterns. (**a**) E-plane. (**b**) H-plane.

3.3. Effects of the Steel Installation Base

Due to the limited mounting space, the characteristics of the proposed antenna were influenced inevitably. In this subsection, the steel installation base was introduced into the simulation model of the antenna in Ansoft HFSS.

3.3.1. Effect on Impedance Characteristics

Figure 9 displays the simulated return loss curves of the proposed antenna with the steel installation base, and other parameters were set as the same values in Figure 5. By comparing the simulated results in Figure 9 and Figure 5, it is observed that when the steel installation base was introduced, the resonant frequencies were obviously increased. Taking the case of R = 56 Ω as an example, the resonant frequency in Figure 5 is 2.40 GHz with a return loss of 30.8 dB, while in Figure 9 that is 2.50 GHz with a return loss of 29.6 dB. Consequently, the addition of the steel installation base further strengthens the inductive nature of the patch antenna and hence increases the operating frequency, which makes it more difficult to realize the required compactness. Another variation between the results of Figure 9 and Figure 5 is the reduced bandwidth. As above mentioned, the simulated bandwidth is 16.3% in the case of R = 56 Ω, while it is reduced to 13.6% from 2.34 to 2.68 GHz due to the influence of the steel installation base.

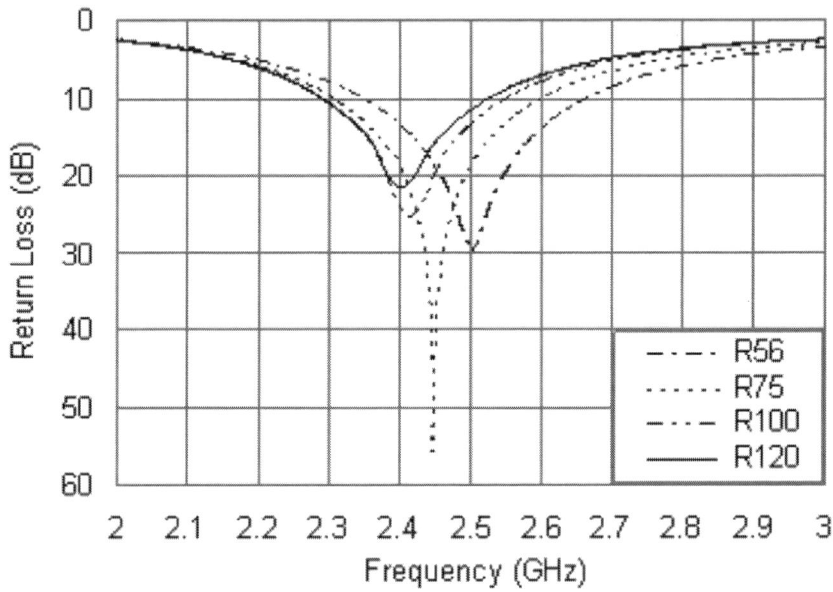

Figure 9. Simulated return loss curves of the proposed antenna along with the installation base.

In order to obtain a good impedance matching for the 2.4 GHz, as can be seen in Figure 9, the resistance of the loading resistor can be improved. It is observed that for R = 120 Ω, the antenna with the steel installation base resonates at 2.4 GHz with a return loss of 22.1 dB, and the bandwidth is 10.4% from 2.28 to 2.53 GHz.

Figure 10 shows the measured and simulated return loss curves of the antenna with the steel installation base for comparison. It is found that the measured resonant frequency is 2.4 GHz with a return loss of 34.2 dB, and the measured bandwidth is 11.7% from 2.27 to 2.55 GHz. Good agreement is achieved between the measured and simulated results. The slight discrepancy in the measured and simulated curves can be attributed to the fabrication tolerances.

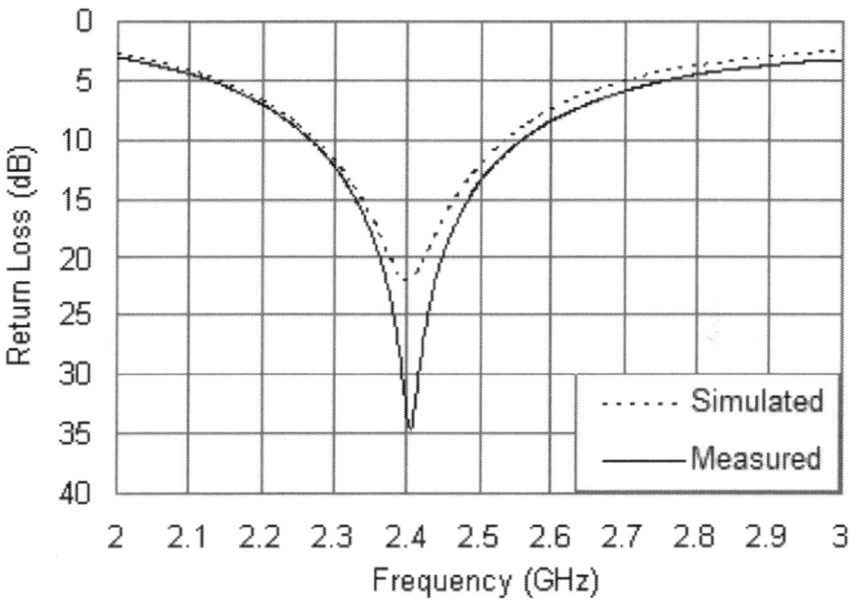

Figure 10. Measured and simulated return loss curves of the antenna along with the installation base.

3.3.2. Effect on Radiation Characteristics

The far-field patterns in both the E- and H-plane were simulated and measured, and the results are shown in Figure 11. In comparison with the results shown in Figure 8, the symmetric characteristics in co-polarization patterns are maintained with the polarization levels slightly reduced. Simultaneously, the backward radiations are weakened in both the planes due to the backed steel installation base, which can be explained as the restraining effects of the metal material. It is also noted that the cross-polarization patterns are changed obviously between Figure 11 and Figure 8, while the polarization levels are remained well.

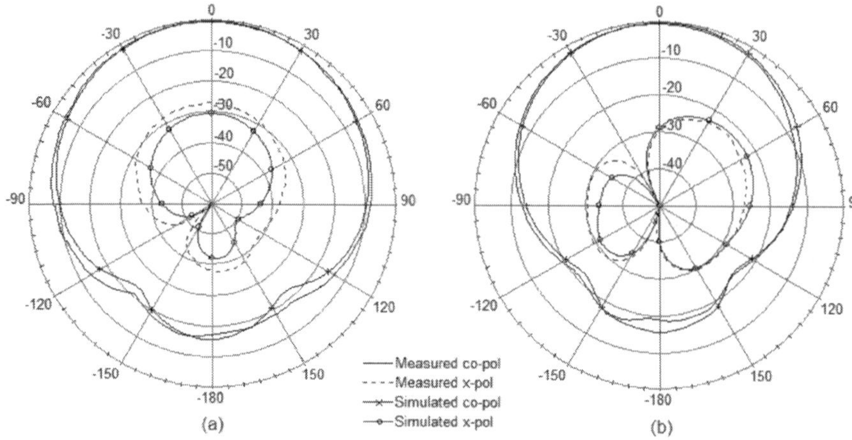

Figure 11. Measured and simulated radiation patterns of the antenna along with the installation base. (**a**) E-plane. (**b**) H-plane.

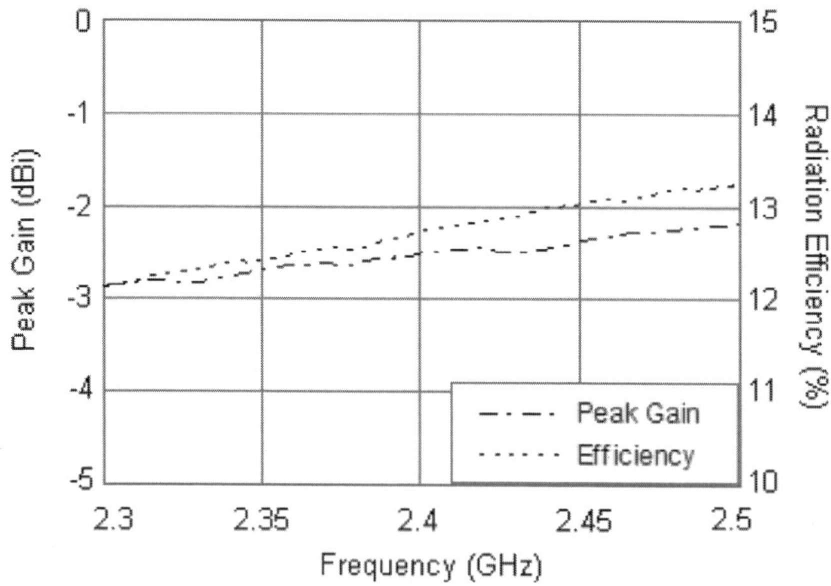

Figure 12. Measured peak gain and radiation efficiency of the antenna along with the installation base.

The measured peak gain of the antenna backed with the installation base is plotted in Figure 12 along with its radiation efficiency. The peak gain is between −2.8 and −2.2 dBi within the operating band (2.3–2.5 GHz) with the radiation efficiency is from 12.2% to 13.3%. For the center frequency of 2.4 GHz, with the influences of installation base and resistor loading, the radiation efficiency of the antenna reduces to 12.7% and the peak gain is - 2.5 dBi.

For further study to improve the gain of the antenna a substrate with relatively low permittivity can be used and its thickness increased suitably, however, the gain improvement may be relatively limited. Another approach to obtain an enhanced antenna gain is by covering the FR4 substrate with a high permittivity superstrate layer, and its thickness needs to be optimized to gain of the required compactness of the antenna. Furthermore, it may be possible to increase the gain by incorporating suitably meandered geometries.

4. CONCLUSIONS

In this paper, a compact annular ring mirostrip antenna backed with a steel installation base was realized. By using a loading resistor of large resistance, 38% miniaturization is achieved compared with a conventional annular ring antenna. It is also found that the steel installation base has a great influence on the antenna's characteristics. With the addition of the steel installation base, the resonant frequency of the antenna increases about 4.2%, while the measured bandwidth was reduced from 17.5% to 11.7% by simultaneously optimizing the load resistance. The realized antenna is well matched at 2.4 GHz with good radiation characteristics. It should also be noted that, due to the dramatic size reduction and influences of the steel installation base, the radiation efficiency reduces from 24.9% to 12.7% and the peak gain is −2.5 dBi. The configuration of proposed antenna is simple and easy to fabricate. Several antennas have been used in the WSN successfully and the performance has been verified. The design provides a reasonable solution to the applicability of microstrip antenna used in narrow spaces.

ACKNOWLEDGMENTS

This work is supported by the Foundation of National Key Laboratory for Electronic Measurement Technology, China (No. 2011YF003).

REFERENCES

1. Bilotti, F.; Alu, A.; Vegni, L. Design of miniaturized metamaterial patch antennas with u-negative loading. IEEE Trans. Antennas Propag. **2008**, 56, 1640–1647.
2. Wong, K.-L.; Huang, C.-H. Printed loop antenna with a perpendicular feed for penta-band mobile phone application. IEEE Trans. Antennas Propag. **2008**, 56, 2138–2141.
3. Chen, N.Z.N. Multipatches multilayered UWB microstrip antennas. IET Microw. Antennas Propag. **2009**, 3, 379–386.
4. Ojaroudi, M.; Kohneshahri, G.; Noory, J. Small modified monopole antenna for UWB application. IET Microw. Antennas Propag. **2009**, 3, 863–869.
5. Kan, H.K.; Pavlickovski, D. Small dual C-slot printed antenna. Electron. Lett. **2003**, 39, 593–594.
6. Bao, X.L.; Ammann, M.J. Microstrip-fed dual-frequency annular-slot antenna loaded by split-ring-slot. IET Microw. Antennas Propag. **2009**, 3, 757–764.
7. Pal, A.; Behera, S.; Vinoy, K.J. Design of multi-frequency microstrip antennas using multiple rings. IET Microw. Antennas Propag. **2009**, 3, 77–84.
8. Sze, J.-Y.; Hsu, C.-I.G.; Hsu, S.-C. Dual-broadband multistandard printed slot antenna with a composite back-patch. IET Microw. Antennas Propag. **2008**, 2, 205–209.
9. Min, C.; Free, C.E. Analysis of circularly polarised dual-ring microstrip patch array using hybrid feed. IET Microw. Antennas Propag. **2009**, 3, 465–472.
10. Ren, Y.-J.; Chang, K. An annual ring antenna for UWB communications. IEEE Antennas Wirel. Propag. Lett. **2006**, 5, 274–276.
11. Krishna, D.D.; Gopikrishna, M.; Aanandan, C.K. Compact wideband Koch fractal printed slot antenna. IET Microw. Antennas Propag. **2009**, 3, 782–789.
12. Hong, W.; Sarabandi, K. Low profile miniaturized planar antenna with omnidirectional vertically polarized radiation. IEEE Trans. Antennas Propag. **2008**, 56, 1533–1540.

13. Zhang, Y.P.; Sun, M.; Lin, W. Novel antenna-in-package design in LTCC for single-chip RF transceivers. IEEE Trans. Antennas Propag. **2008**, 56, 2079–2088.

14. Waterhouse, R.B.; Targonski, S.D.; Kokotoff, D.M. Design and performance of small printed antennas. IEEE Trans. Antennas Propag. **1998**, 46, 1629–1633.

15. Kan, H.K.; Waterhouse, R.B. Size reduction technique for shorted patches. Electron. Lett. **1999**, 35, 948–949.

16. Sim, C.-Y.-D.; Tu, S.-Y. Dual frequency shorted patch antenna for WLAN application. Microw. Opt. Tech. Lett. **2007**, 49, 389–391.

17. Yoon, J.H. Fabrication and measurement of a compact meander-line-type microstrip patch antenna with shorting pin for the 5-GHz band. Microw. Opt. Tech. Lett. **2006**, 48, 944–947.

18. Wang, Y.J.; Lee, C.K.; Koh, W.J. Single-patch and single-layer square microstrip antenna with 67.5% bandwidth. IEE Proc. Microw. Antennas Propag. **2001**, 148, 418–422.

19. Kaifas, T.N.; Sahalos, J.N. Analysis of printed antennas mounted on a coated circular cylinder of arbitrary size. IEEE Trans. Antennas Propag. **2006**, 54, 2797–2807.

20. Yuan, N.; Yeo, T.S.; Nie, X. -C. Analysis of probe-fed conformal microstrip antennas on finite grounded substrate. IEEE Trans. Antennas Propag. **2006**, 54, 554–563.

21. Kokotoff, D.M.; Aberle, J.T.; Waterhouse, R.B. Rigorous analysis of probe-fed printed annular ring antennas. IEEE Trans. Antennas Propag. **1999**, 47, 384–388.

22. Thomas, K.G.; Sreenivasan, M. A simple dual-band microstrip-fed printed antenna for WLAN applications. IET Microw. Antennas Propag. **2009**, 3, 687–694.

23. Mahajan, M.; Khah, S.K.; Chakarvarty, T. Computation of resonant frequency of annular microstrip antenna loaded with multiple shorting posts. IET Microw. Antennas Propag. **2008**, 2, 1–5.

24. Wong, K.-L.; Hsieh, G.-B. Broadband circularly polarized microstrip antenna with a chip-resistor loading. Microw. Opt. Tech. Lett. **1998**, 19, 34–36.

25. Wu, J.-Y.; Huang, C.-Y.; Wong, K.-L. Compact broadband circularly polarized square microstrip antenna. Microw. Opt. Tech. Lett. **1999**, 21, 423–425.

26. Wong, K.-L.; Lin, Y.-F. Microstrip-line-fed compact broadband circular microstrip antenna with chip-resistor loading. Microw. Opt. Tech. Lett. **1998**, 17, 53–55.

27. Wong, K.-L.; Lin, Y.-F. Small broadband rectangular microstrip antenna with chip-resistor loading. Electron. Lett. **1997**, 33, 1593–1594.

7

CHAPTER

WEARABLE ANTENNAS FOR MEDICAL APPLICATIONS

Albert Sabban[1, 2, 3]

[1] Ort Braude College, Karmiel, Israel

[2] Tel Aviv University, Israel

[3] Colorado University, Boulder, USA

1. INTRODUCTION

Microstrip antennas are widely employed in communication system and seekers. Microstrip antennas posse's attractive features such as low profile, flexible, light weight, small volume and low production cost. In addition, the benefit of a compact low cost feed network is attained by integrating the RF frontend with the radiating elements on the same substrate. Microstrip antennas are widely presented in books and papers in the last decade [1-7]. However, the effect of human body on the electrical performance of wearable antennas at 434 MHz is not presented [8-13]. RF transmission properties of human tissues have been investigated in several articles [8-9]. Several wearable antennas have been presented in the last decade [10-14]. A review of wearable and body mounted antennas designed and developed for various applications at different frequency bands over the last decade can be found in [10]. In [11] meander wearable antennas in close proximity of a human body are presented in the frequency range between 800 MHz and 2700 MHz. In [12] a textile antenna performance in the vicinity of the human body is presented at 2.4 GHz. In [13] the effect of human body on wearable 100 MHz portable radio antennas is studied. In [13] the authors concluded that wearable antennas need to be shorter by 15% to 25% from the antenna length in free-space. Measurement of the antenna gain in [13] shows that a wide dipole (116 x 10 cm) has -13dBi gain. The antennas presented in [10-13] were developed mostly for cellular applications. Requirements and the frequency range for medical applications are different from those for cellular applications

In this chapter, a new class of wideband compact wearable microstrip antennas for medical applications is presented. Numerical results with and without the presence of the human body are discussed. The antennas VSWR is better than 2:1at 434 MHz + 5%. The antenna beam width is around 100º. The antennas gain is around 0 to 4 dBi. The antenna resonant frequency is shifted by 5% if the air spacing between the antenna and the human body is increased from 0 mm to 5 mm.

2. DUALLY POLARIZED 434 MHZ PRINTED ANTENNA

A new compact microstrip loaded dipole antennas has been designed to provide horizontal polarization. The antenna dimensions have been optimized to operate on the human body by employing Agilent Advanced Design System (ADS) software [16]. The antenna consists of two layers. The first layer consists of RO3035 0.8 mm dielectric substrate. The second layer consists of RT-Duroid 5880 0.8 mm dielectric substrate. The substrate thickness determines the antenna bandwidth. However, thinner antennas are flexible. Thicker antennas have been designed with wider bandwidth. The printed slot antenna provides a vertical polarization. In several medical systems the required polarization may be vertical or horizontal. The proposed antenna is dually polarized. The printed dipole and the slot antenna provide dual orthogonal polarizations. The dimensions of the dual polarized antenna presented in Figure 1are 26 x 6 x 0.16 cm. The antenna may be used as a wearable antenna on a human body. The antenna may be attached to the patient shirt, patient stomach, or in the back zone. The antenna has been analyzed by using Agilent ADS software. There is a good agreement between measured and computed results. The antenna bandwidth is around 10% for VSWR better than 2:1. The antenna beam width is around 100º. The antenna gain is around 2 dBi. The computed S_{11} and S_{22} parameters are presented in Figure 2. Figure 3 presents the antenna measured S_{11} parameters. The computed radiation patterns are shown in Figure 4. The co-polar radiation pattern belongs to the yz plane. The cross-polar radiation pattern belongs to the xz plane. The antenna cross polarized field strength may be adjusted by varying the slot feed location. The dimensions of the folded dually polarized antenna presented in Figure 5 are 7 x 5 x 0.16 cm. Figure 6 presents the antenna computed S_{11} and S_{22} parameters. The computed radiation patterns of the folded dipole are shown in Figure 7. The antennas radiation characteristics on human body have been measured by using a phantom. The phantom electrical characteristics represent the human body electrical characteristics.

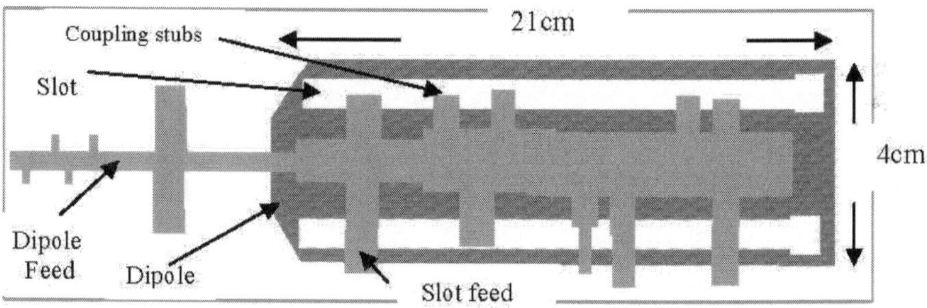

Figure 1. Printed dually polarized antenna, 26 x 6 x 0.16 cm.

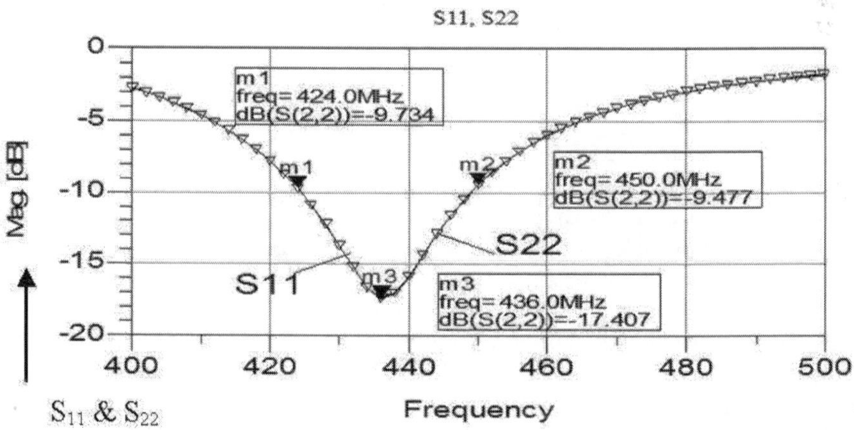

Figure 2. Computed S_{11} and S_{22} results

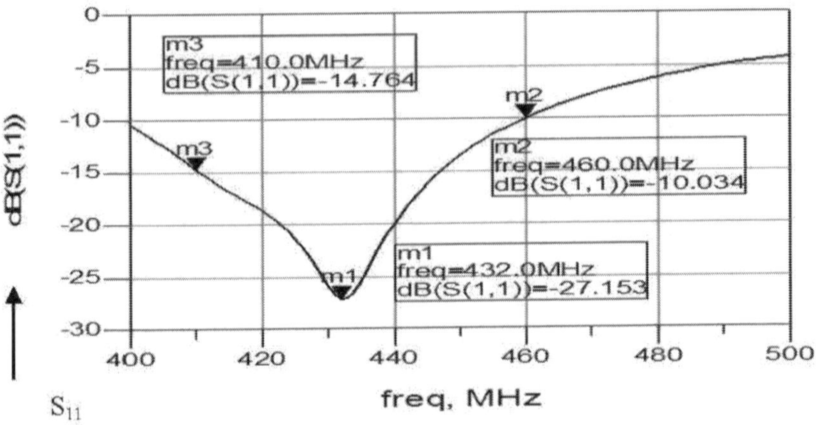

Figure 3. Measured S_{11} on human body

The phantom has a cylindrical shape with a 40cm diameter and a length of 1.5m. The phantom contains a mix of 55% water 44% sugar and 1% salt. The antenna under test was placed on the phantom during the measurements of the antennas radiation characteristics. S_{11} and S_{12} parameters were measured directly on human body by using a network analyzer. The measured results were compared to a known reference antenna.

3. NEW LOOP ANTENNA WITH GROUND PLANE

A new loop antenna with ground plane has been designed on Kapton substrates with thickness of 0.25mm and 0.4mm. The antenna without ground plane is shown in Figure 8. The loop antenna VSWR without the tuning capacitor was 4:1. This loop antenna may be tuned by adding a capacitor or varactor as shown in Figure 8. Tuning the antenna allow us to work in a wider bandwidth. Figure 9 presents the loop antenna computed S_{11}on human body. There is good agreement between measured and computed S_{11}. The computed radiation pattern is shown in Fig 10.

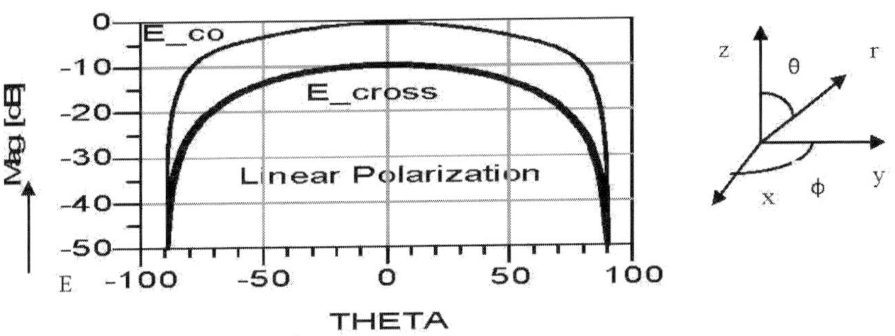

Figure 4. Antenna Radiation patterns

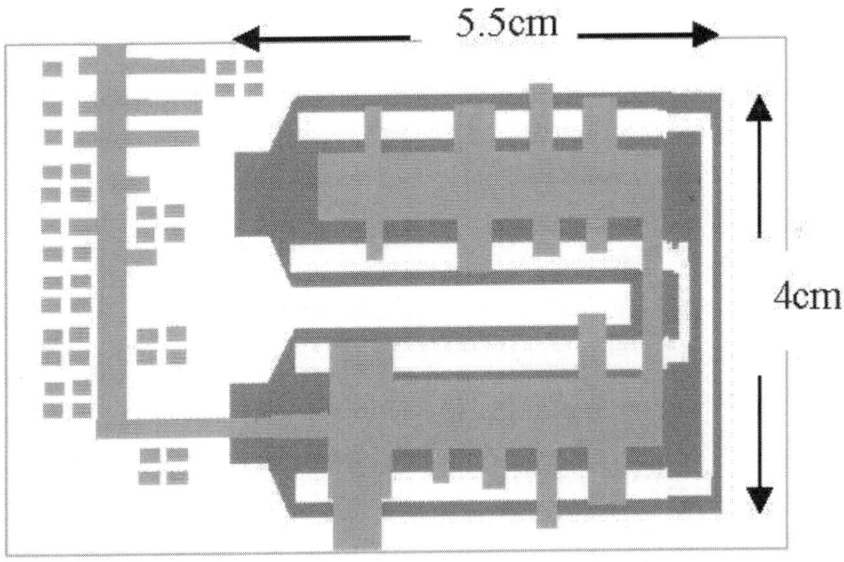

Figure 5. Folded dual polarized antenna, 7x5x0.16cm.

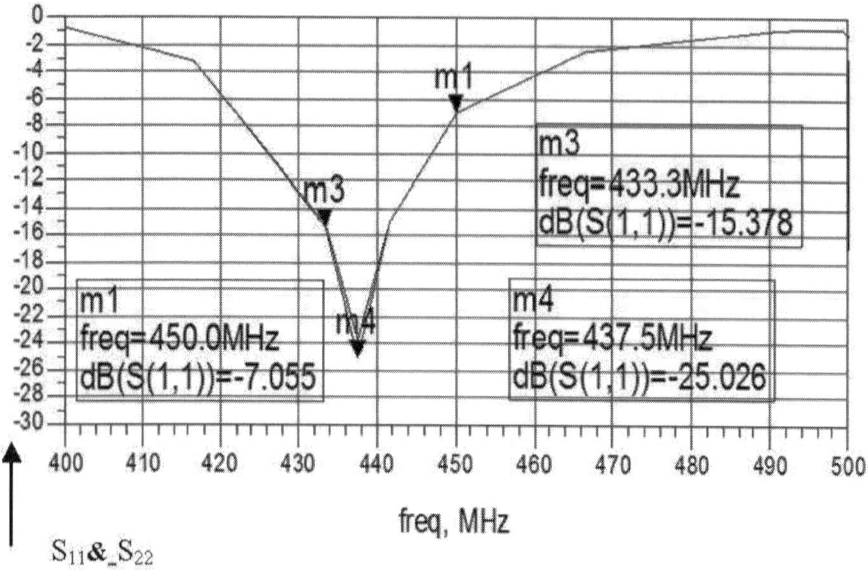

Figure 6. Folded antenna Computed S_{11} and S_{22} results

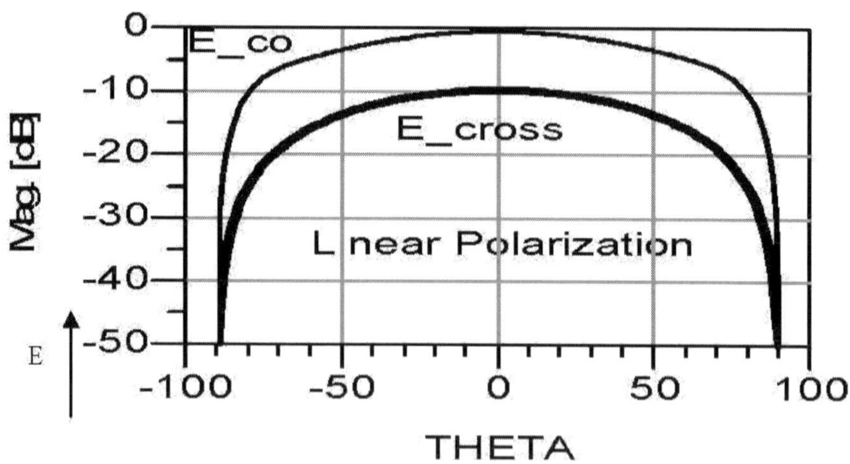

Figure 7. Folded antenna Radiation patterns

Table 1. Comparison of Loop Antennas

Antenna with no tuning capacitor	Beam width 3dB	Gain dBi	VSWR
Loop no GND	100°	0	4:1
Loop with GND	100°	0	2:1

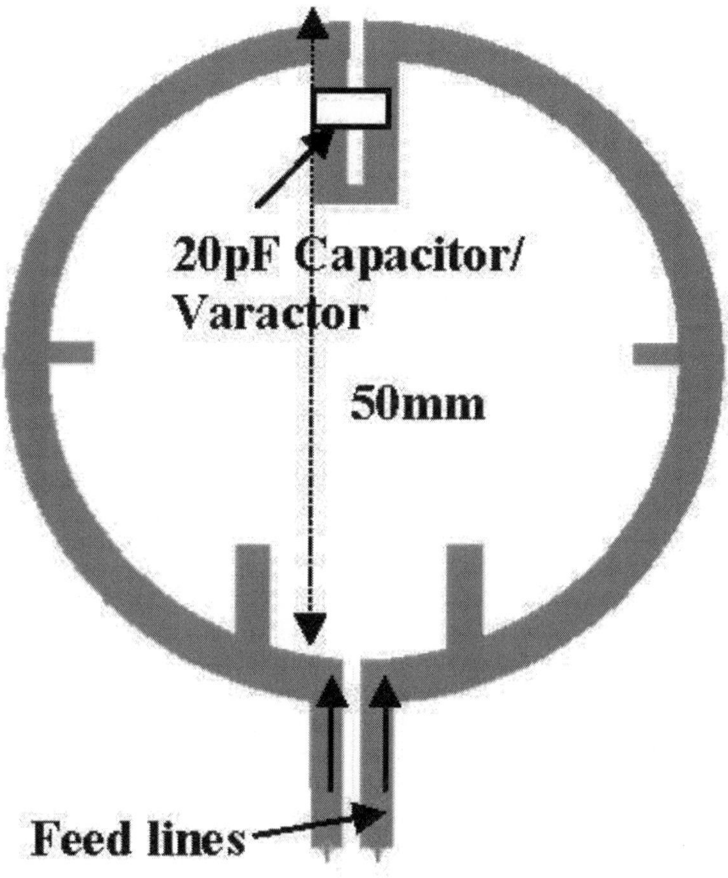

Figure 8. Tunable loop antenna without ground plane

Table I compares the electrical performance of a loop antenna with ground plane with a loop antenna without ground plane. Tuning the antenna allow us to work in a wider bandwidth. Figure 9 presents the loop antenna computed S11on human body.

There is good agreement between measured and computed S_{11}. The computed radiation pattern is shown in Fig 10. Table I compares the electrical performance of a loop antenna with ground plane with a loop antenna without ground plane. There is a good agreement between measured and computed results of antenna parameters on human body. The results presented in Table I indicates that the loop antenna with ground plane is matched to the human body environment, without the tuning capacitor, better than the loop antenna without ground plane. The computed 3D radiation pattern is shown in Fig 11.

The computed 3D radiation pattern and the coordinate used in this chapter are shown in Fig 11. Computed S_{11} of the Loop Antenna with a tuning capacitor is given in Figure 12.

4. ANTENNA S11 VARIATION AS FUNCTION OF DISTANCE FROM BODY

The Antennas input impedance variation as function of distance from the body had been computed by employing ADS software. The analyzed structure is presented in Figure 14. The patient body thickness was varied from 15mm to 300mm. The dielectric constant of the body was varied from 40 to 50. The antenna was placed inside a belt with thickness between 2 to 4mm with dielectric constant from 2 to 4. The air layer between the belt and the patient shirt may vary from 0mm to 8mm. The shirt thickness was varied from 0.5mm to 1mm. The dielectric constant of the shirt was varied from 2 to 4. Properties of human body tissues are listed in Table II see [8]. These properties were employed in the antenna design. Figure 15 presents S_{11} results (of the antenna shown in Figure 1) for different belt thickness, shirt thickness and air spacing between the antennas and human body.

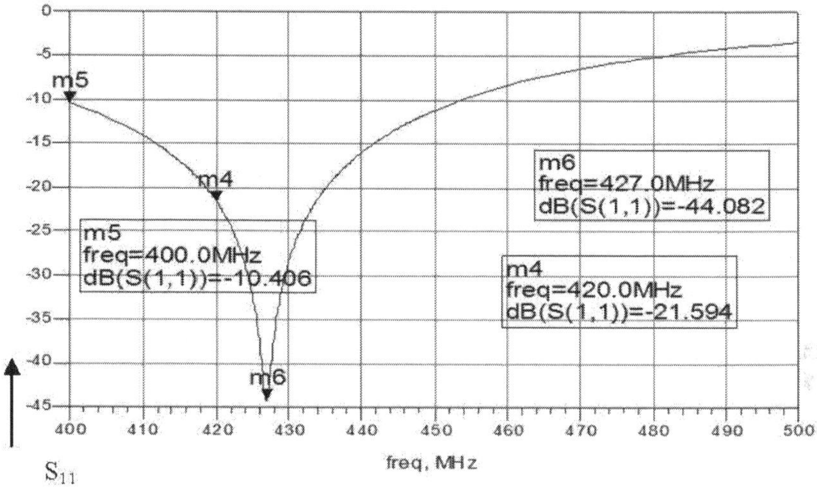

Figure 9. Computed S_{11} of new Loop Antenna

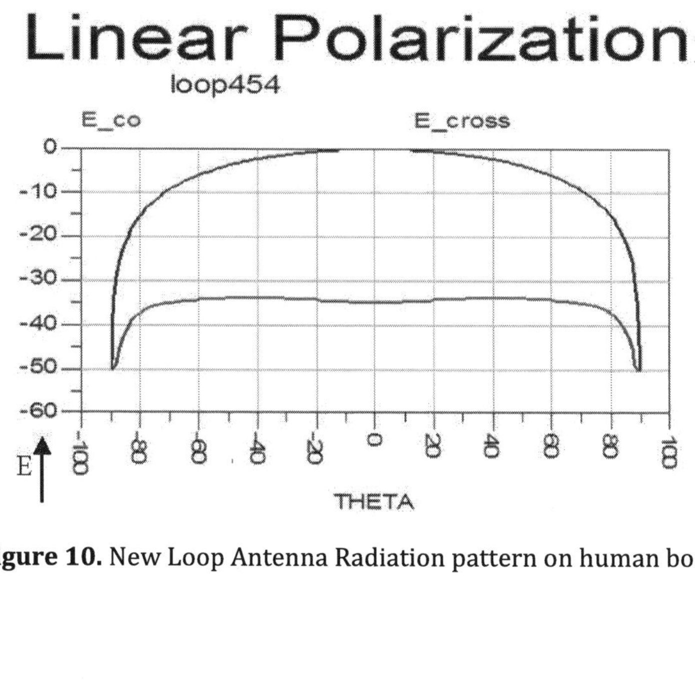

Figure 10. New Loop Antenna Radiation pattern on human body

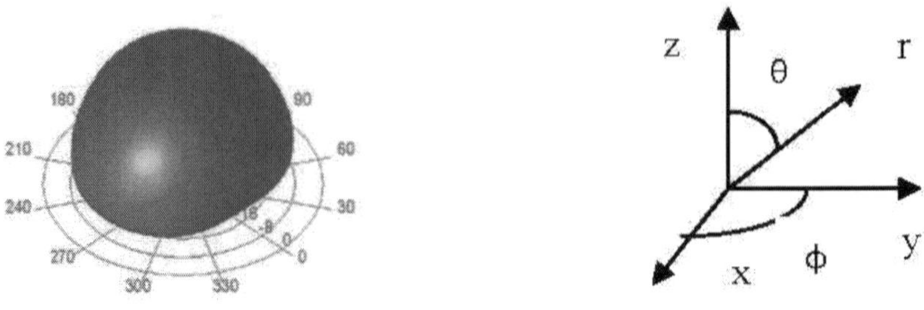

Figure 11. New Loop Antenna 3D Radiation pattern

Table 2. Properties of human body tissues

Tissue	Property	434 MHz	600 MHz
Skin	σ	0.57	0.6
	ε	41.6	40.43
Stomach	σ	0.67	0.73
	ε	42.9	41.41
Colon, Muscle	σ	0.98	1.06
	ε	63.6	61.9
Lung	σ	0.27	0.27
	ε	38.4	38.4

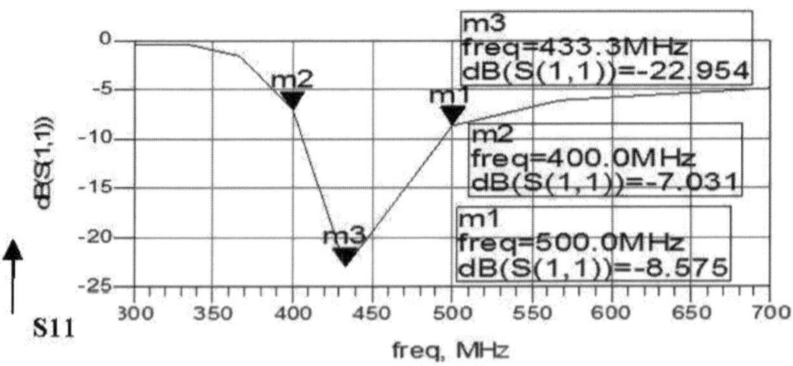

Figure 12. Computed S_{11} of Loop Antenna, without ground plane, with a tuning capacitor

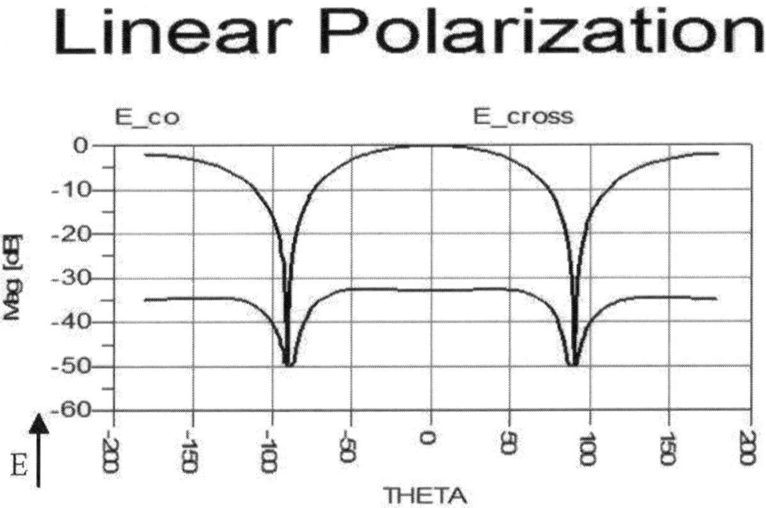

Figure 13. Radiation pattern of Loop Antenna without ground on human body

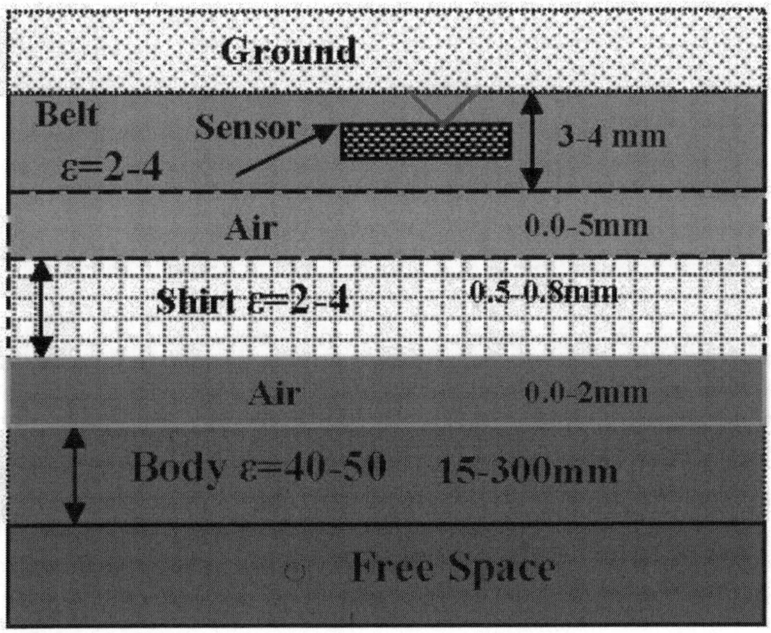

Figure 14. Analyzed structure for Impedance calculations

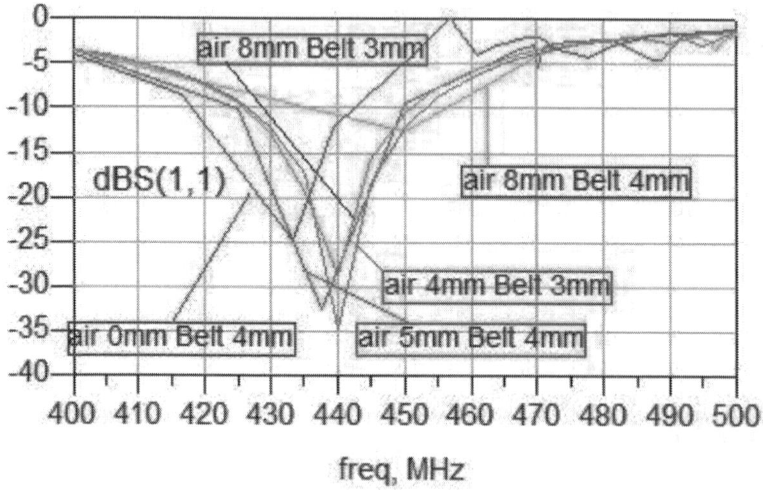

Figure 15. S_{11} results for different antenna positions relative to the human body

One may conclude from results shown in Figure 15 that the antenna has V.S.W.R better than 2.5:1 for air spacing up to 8mm between the antennas and patient body. For frequencies ranging from 415MHz to 445MHz the antenna has V.S.W.R better than 2:1 when there is no air spacing between the antenna and the patient body. Results shown in Figure 16 indicates that the folded antenna (the antenna shown in Figure 5) has V.S.W.R better than 2.0:1 for air spacing up to 5mm between the antennas and patient body. Figure 16 presents S_{11} results of the folded antenna results for different position relative to the human body. Explanation of Figure 16 is given in Table 3. If the air spacing between the sensors and the human body is increased from 0mm to 5mm the antenna resonant frequency is shifted by 5%. The loop antenna with ground plane has V.S.W.R better than 2.0:1 for air spacing up to 5mm between the antennas and patient body.

If the air spacing between the sensors and the human body is increased from 0mm to 5mm the computed antenna resonant frequency is shifted by 2%. However, if the air spacing between the sensors and the human body is increased up to 5mm the measured loop antenna resonant frequency is shifted by 5%. Explanation of Figure 17 is given in Table 4.

Table 3. Explanation of Figure 16

Plot colure	Sensor position
Red	Shirt thickness 0.5mm
Blue	Shirt thickness 1mm
Pink	Air spacing 2mm
Green	Air spacing 4mm
Sky	Air spacing 1mm
Purple	Air spacing 5mm

5. WEARABLE ANTENNA

An application of the proposed antenna is shown in Figure 18. Three to four folded dipole or loop antennas may be assembled in a belt and attached to the patient stomach. The cable from each antenna is connected to a recorder. The received signal is routed to a switching matrix. The signal with the highest level is selected during the medical test. The antennas receive a signal that is transmitted from various positions in the human body. Folded antennas may be also attached on the patient back in order to improve the level of the received signal from different locations in the human body. Figure 19 and Figure 20 show various antenna locations on the back and front of the human body for different medical applications.

Table 4. Explanation of Figure 17

Plot colure	Sensor position
Red	Body 15mm air spacing 0mm
Blue	Air spacing 5mm Body 15mm
Pink	Body 40mm air spacing 0mm
Green	Body 30mm air spacing 0mm
Sky	Body 15mm Air spacing 2mm
Purple	Body 15mm Air spacing 4mm

transmitting and receiving antennas is less than $2D^2/\lambda$. D is the largest dimension of the radiator. In these applications the amplitude of the electromagnetic field close to the antenna may be quite powerful, but because of rapid fall-off with distance, the antenna do not radiate energy to

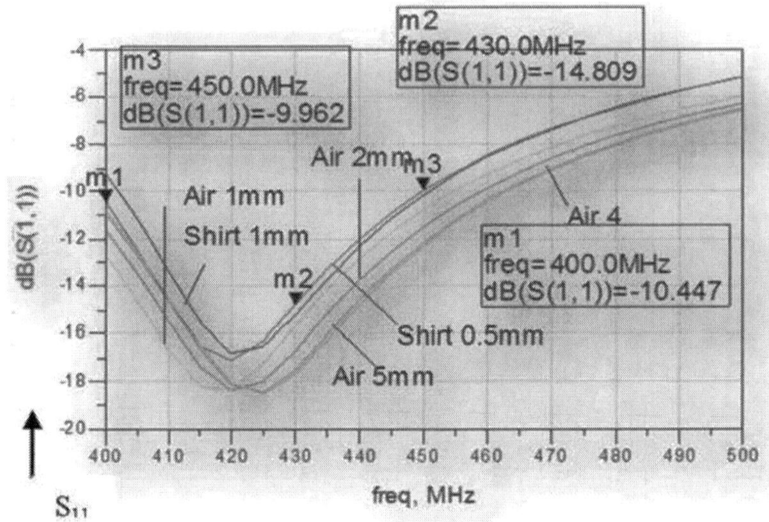

Figure 16. Folded antenna S_{11} results for different antenna position relative to the human body

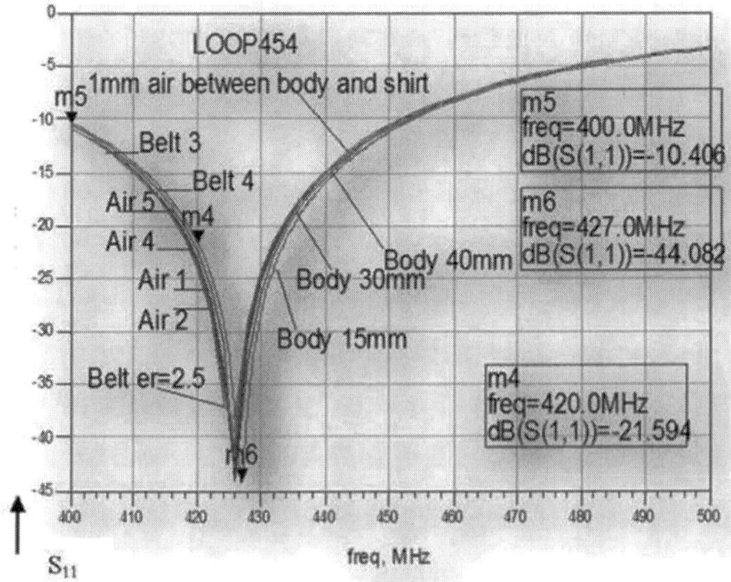

Figure 17. Loop antenna S_{11} results for different antenna position relative to the human body

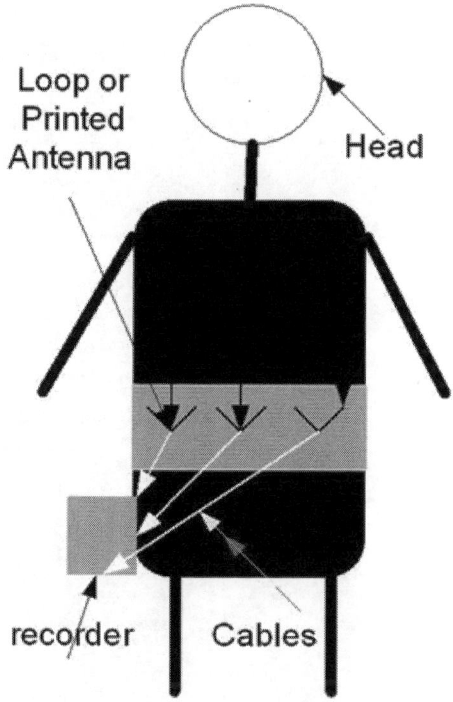

Figure 18. Printed Wearable antenna

infinite distances, but instead the radiated power remain trapped in the region near to the antenna. Thus, the near-fields only transfer energy to close distances from the receivers. The receiving and transmitting antennas are magnetically coupled. Change in current flow through one wire induces a voltage across the ends of the other wire through electromagnetic induction. The amount of inductive coupling between two conductors is measured by their mutual inductance. In these applications we have to refer to the near field and not to the far field radiation.

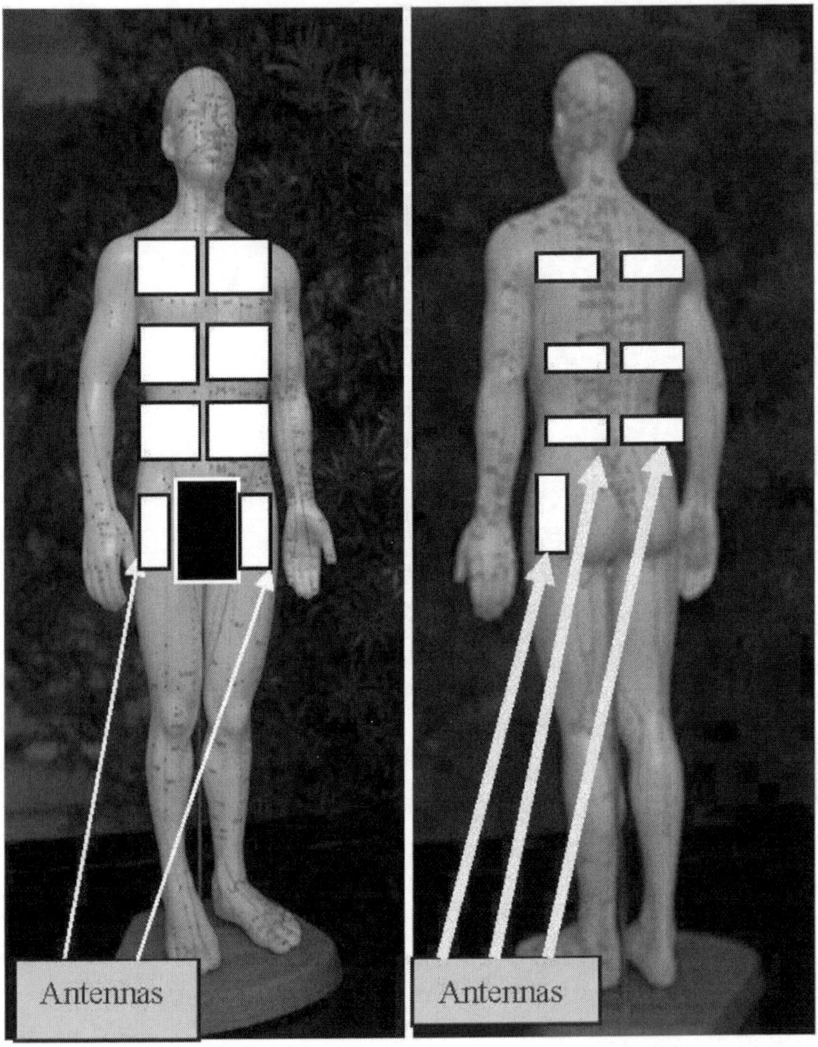

Figure 19. Printed Patch Antenna locations for various medical applications

In Figure 20 and 21 several microstrip antennas for medical applications at 434MHz are shown. The Backside of the antennas is presented in Figure 20.b. The diameter of the loop antenna presented in Figure 21 is 50 mm. The dimensions of the folded dipole antenna are 7x6x0.16cm. The dimensions of the compact folded dipole presented in Figure 21 are 5x5x0.5cm.

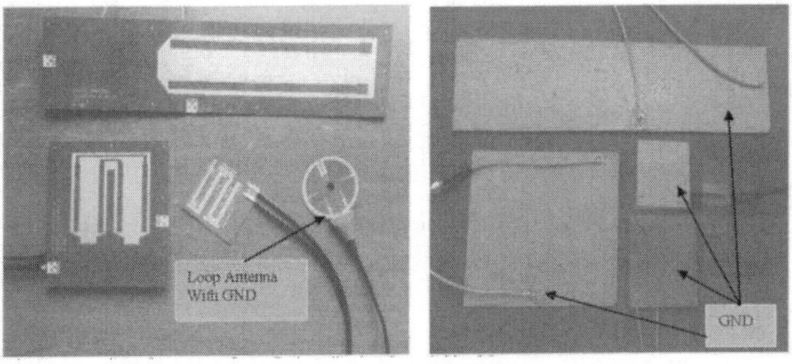

Figure 20. a. Microstrip Antennas for medical Applications b. Backside of the antennas6. Compact dual polarized printed antenna

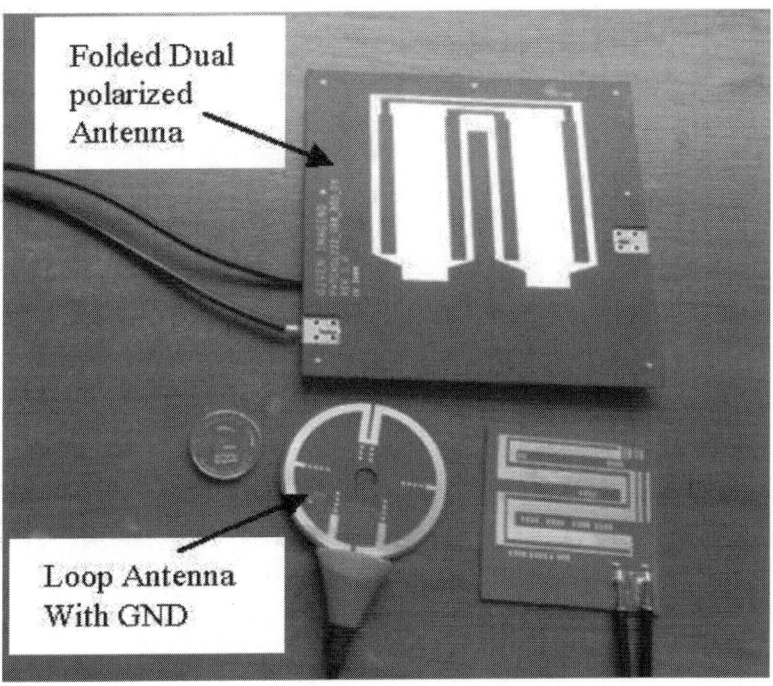

Figure 21. Microstrip Antennas for medical Applications

A new compact microstrip loaded dipole antennas has been designed. The antenna consists of two layers. The first layer consistsof FR4 0.25mm dielectric substrate. The second layer consists of Kapton 0.25mm dielectric substrate. The substrate thickness determines the antenna bandwidth. However, with thinner substrate we may achieve better flexibility. The proposed antenna is dual polarized. The printed dipole and the slot antenna provide dual orthogonal polarizations. The dual polarized antenna is shown in Figure 22. The antenna dimensions are 5x5x0.05cm.

The antenna may be attached to the patient shirt in the patient stomach or back zone. The antenna has been analyzed by using Agilent ADS software. There is a good agreement between measured and computed results. The antenna bandwidth is around 10% for VSWR better than 2:1. The antenna beam width is around 100º. The antenna gain is around 0dBi. The computed S_{11} parameters are presented in Figure 23. Figure 24 presents the antenna measured S_{11} parameters. The antenna cross-polarized field strength may be adjusted by varying the slot feed location. The computed 3D radiation pattern of the antenna is shown in Figure 25. The computed radiation pattern is shown in Figure 26.

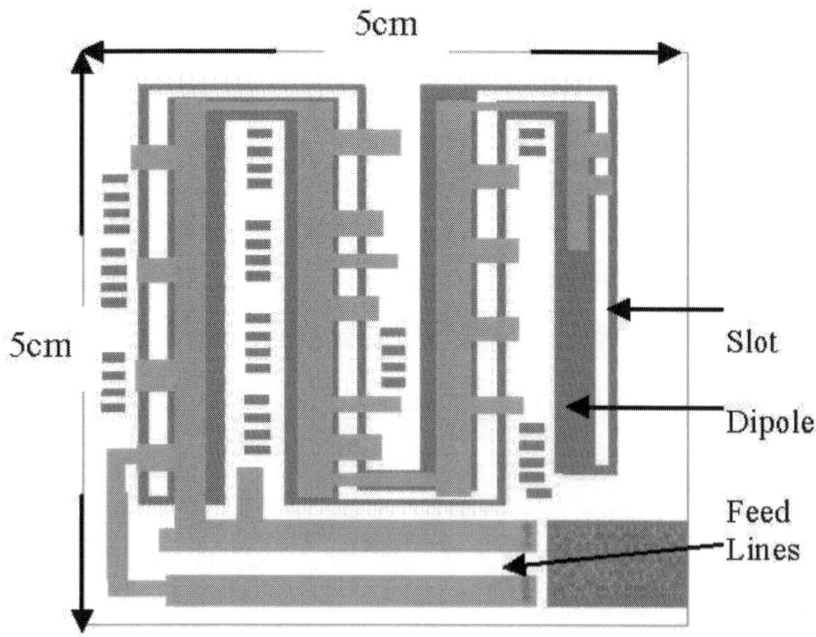

Figure 22. Printed Compact dual polarized antenna

6. HELIX ANTENNA PERFORMANCE ON HUMAN BODY

In order to compare the variation of the new antenna input impedance as function of distance from the body to other antennas a helix antenna has been designed. A helix antenna with 9 turns is shown in figure 27. The backside of the circuit is copper under the microstrip matching stubs. However, in the helix antenna area there is no ground plane. The antenna has been designed to operate on human body. A matching microstrip line network has been designed on RO4003 substrate with 0.8mm thickness. The helix antenna has VSWR better than 3:1 at the frequency range from 440MHz to 460MHz. The antenna dimensions are 4x4x0.6cm. Figure 28 presents the measured S_{11} parameters on human body. The computed E and H radiation plane of the helix antenna is shown in Figure 29. The helix antenna input impedance variation as function of distance from the body is very sensitive. If the air spacing between the helix antenna and the human body is increased from 0mm to 2mm the antenna resonant frequency is shifted by 5%.

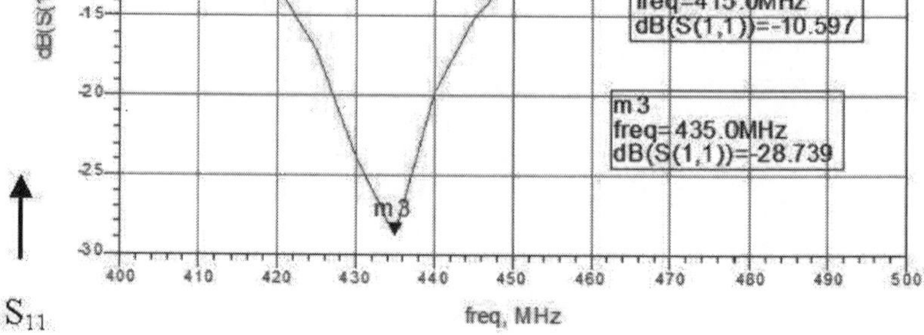

Figure 23. Computed S_{11} results of compact antenna

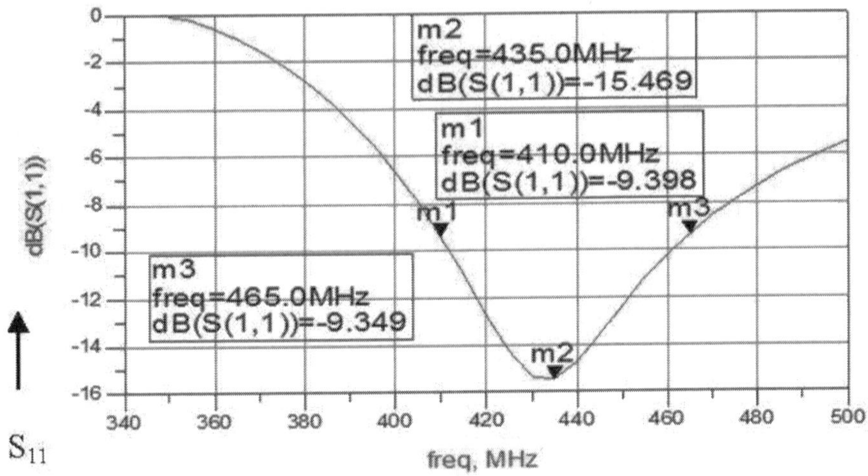

Figure 24. Measured S_{11} on human body

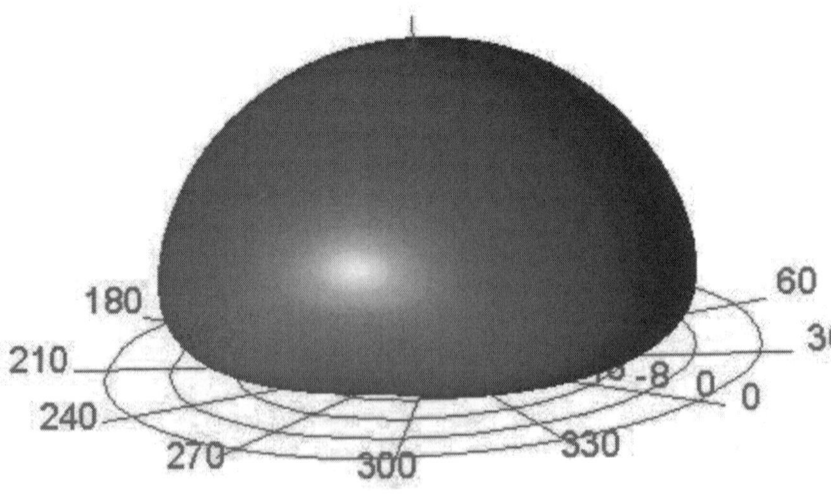

Figure 25. Compact Antenna 3D Radiation pattern

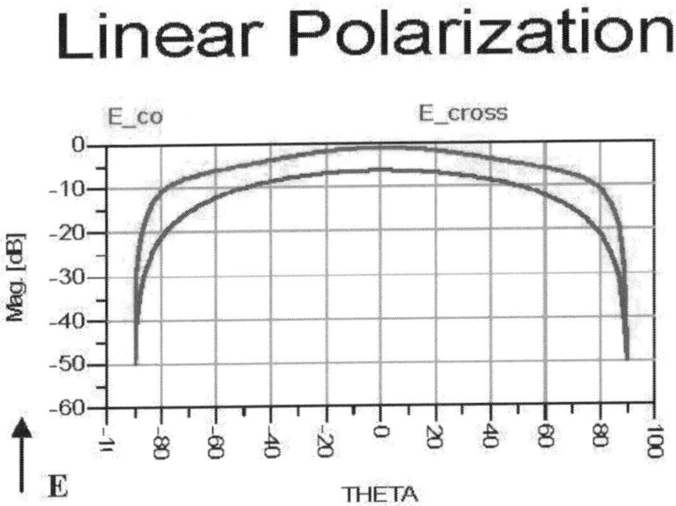

Figure 26. Antenna Radiation pattern

Figure 27. Helix Antenna for medical Applications

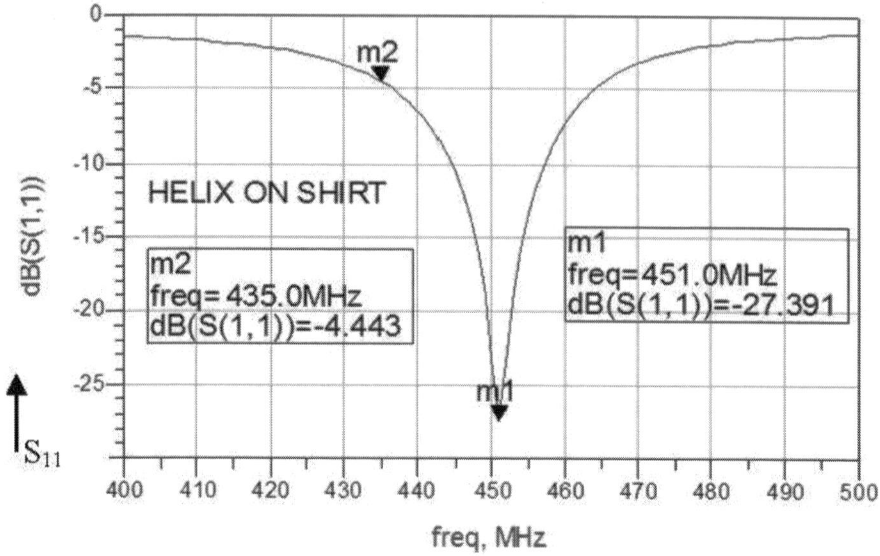

Figure 28. Measured S_{11} on human body

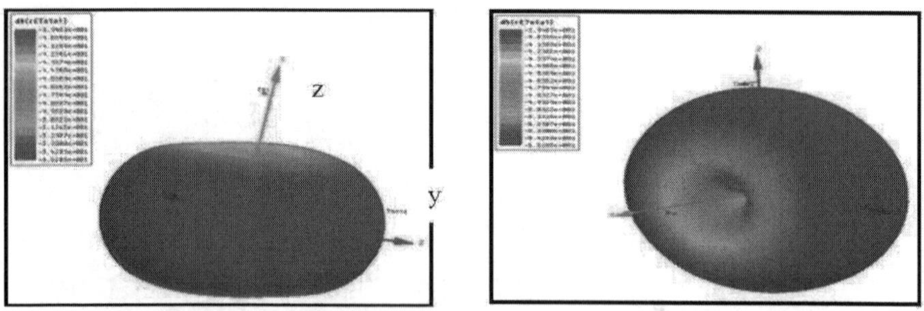

Figure 29. E and H plane radiation pattern of The Helix antenna

However, if the air spacing between the new dual polarized antenna and the human body is increased from 0mm to 5mm the antenna resonant frequency is shifted only by 5%.

7. WEARABLE TUNABLE PRINTED ANTENNAS FOR MEDICAL APPLICATIONS

A new class of wideband tunable wearable microstrip antennas for medical applications is presented in this section. The antennas VSWR is better than 2:1at 434MHz+5%. The antenna beam width is around 100º. The antennas gain is around 0 to 2dBi. A voltage controlled varactor is used to control the antenna resonant frequency at different locations on the human body.

7.1. Dually polarized tunable printed antenna

A compact tunable microstrip dipole antenna has been designed to provide horizontal polarization. The antenna consists of two layers. The first layer consists of RO3035 0.8mm dielectric substrate. The second layer consists of RT-Duroid 5880 0.8mm dielectric substrate. The substrate thickness affects the antenna band width. The printed slot antenna provides a vertical polarization. The printed dipole and the slot antenna provide dual orthogonal polarizations. The dimensions of the dual polarized antenna are 26x6x0.16cm. Also tunable compact folded dual polarized antennas have been designed. The dimensions of the compact antennas are 5x5x0.05cm.Varactors are connected to the antenna feed lines as shown in Figure 30. The voltage controlled varactors are used to control the antenna resonant frequency. The varactor bias voltage may be varied automatically to set the antenna resonant frequency at different locations on the human body. The antenna may be used as a wearable antenna on a human body. The antenna may be attached to the patient shirt in the patient stomach or back zone. The antenna has been analyzed by using Agilent ADS software. There is a good agreement between measured and computed results. The antenna bandwidth is around 10% for VSWR better than 2:1. The antenna beam width is around 100º. The antenna gain is around 2dBi.

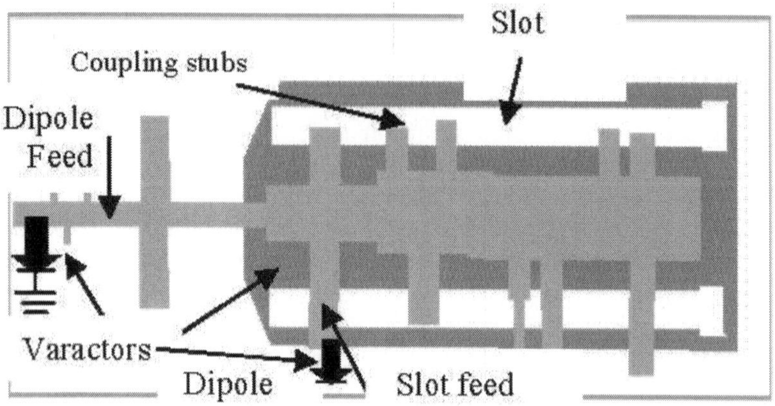

Figure 30. Dual polarized tunable antenna, 26x6x0.16 cm.

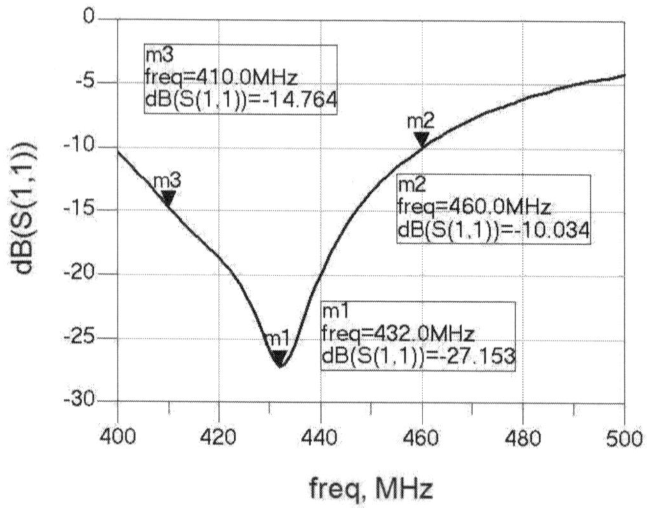

Figure 31. Measured S_{11} on human body

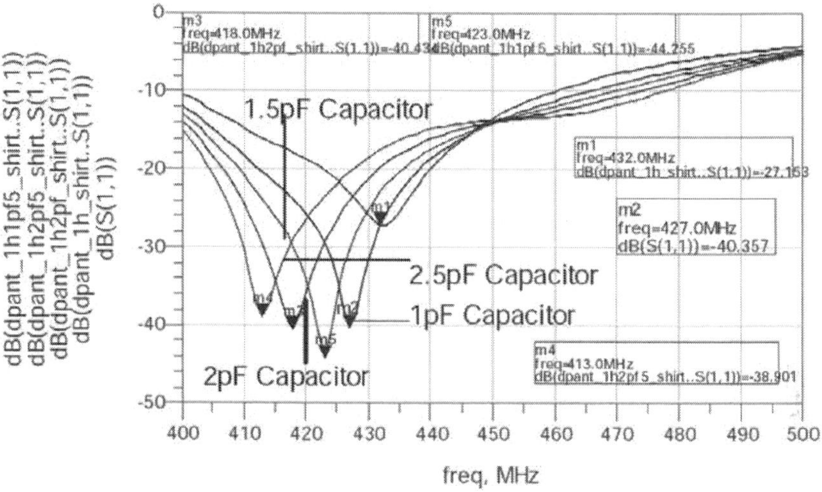

Figure 32. The Tunable S_{11} parameter as function of varactor capacitance

Figure 31 presents the antenna measured S_{11} parameters without a varactor. Figure 32 presents the antenna S_{11} parameters as function of different varactor capacitances. Figure 33 presents the tunable antenna resonant frequency as function of the varactor capacitance. The antenna resonant frequency varies around 5% for capacitances up to 2.5pF. The antenna beam width is 100º. The antenna cross polarized field strength may be adjusted by varying the slot feed location.

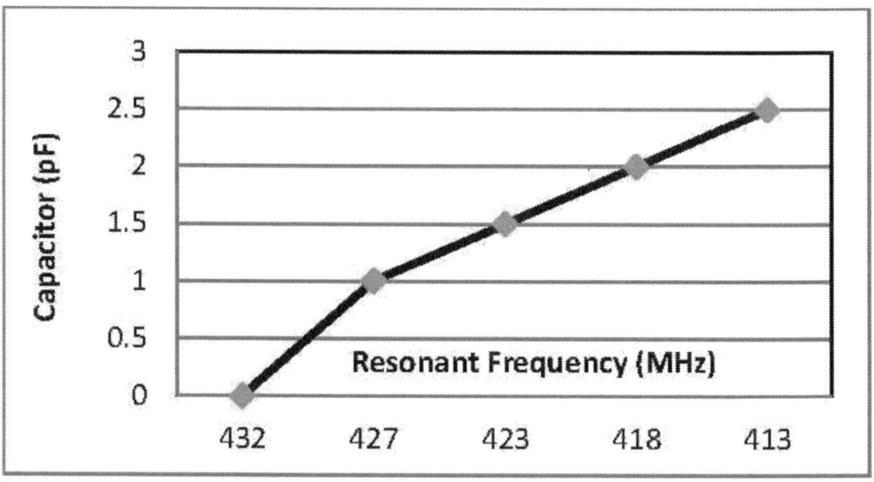

Figure 33. Resonant frequency as function of varactor capacitance

7.2. Antenna S11 varitaion as function of distance from body

The Antennas input impedance variation as function of distance from the body had been computed by employing ADS software. The analyzed structure is presented in Figure 14. Properties of human body tissues are listed in Table 2 see [8]. Figure 34 presents S_{11} results for different belt and shirt thickness, and air spacing between the antennas and human body.

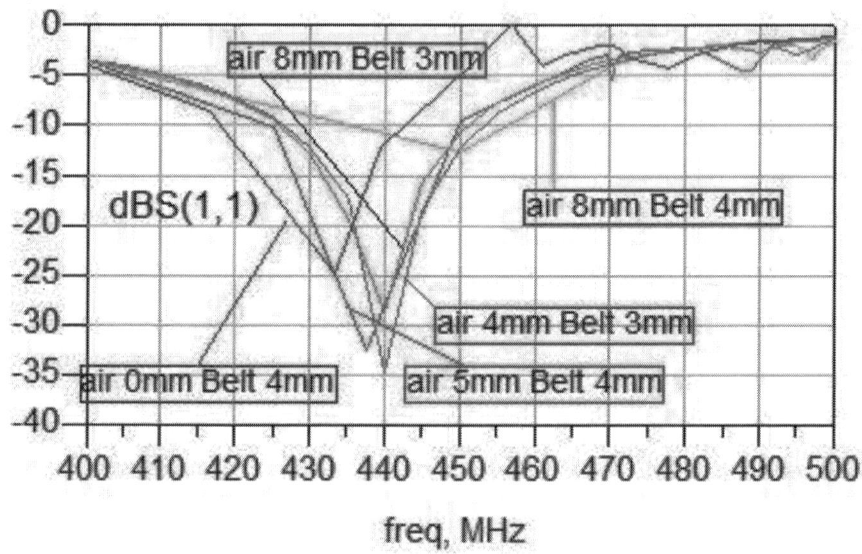

Figure 34. S_{11} results for different antenna positions

If the air spacing between the sensors and the human body is increased from 0mm to 5mm the antenna resonant frequency is shifted by 5%. There is good agreement between measured and calculated results. The voltage controlled varactor may be used to tune the antenna resonant frequency due to different antenna locations on a human body. Figure 35 presents several compact tunable Antennas for medical Applications. A voltage controlled varactor may be used also to tune the loop antenna resonant frequency at different antenna locations on the body.

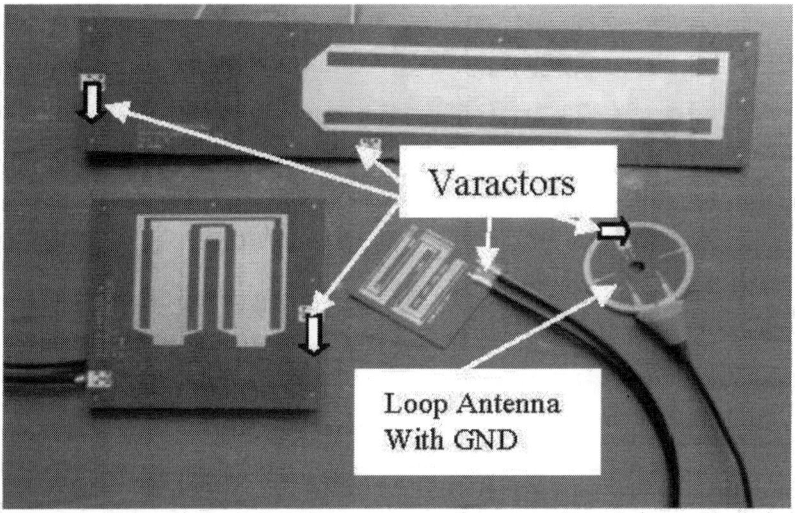

Figure 35. Tunable Antennas for medical Applications

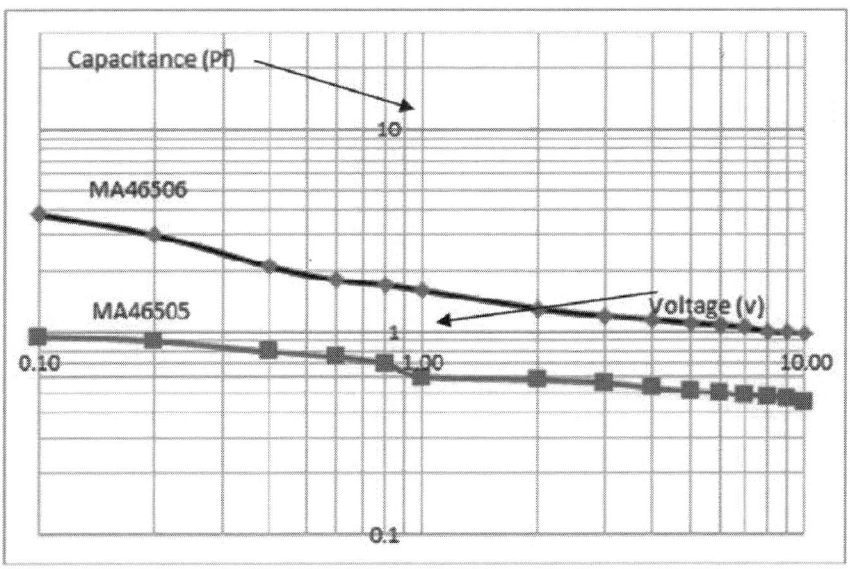

Figure 36. Varactor capacitance as function of bias voltage

7.3. Varactors

Tuning varactors are voltage variable capacitors designed to provide electronic tuning of microwave components. Varactors are manufactured on silicon and gallium arsenide substrates. Gallium arsenide varactors offer higher Q and may be used at higher frequencies than silicon varactors. Hyperabrupt varactors provide nearly linear variation of frequency with applied control voltage. However abrupt varactors provide inverse fourth root frequency dependence. MACOM offers several gallium arsenide hyperabrupt varactors such as MA46 series. Figure 36 presents the C-V curves of varactors MA46505 to MA46506.

Figure 37 presents the C-V curves of varactors MA46H070 to MA46H074

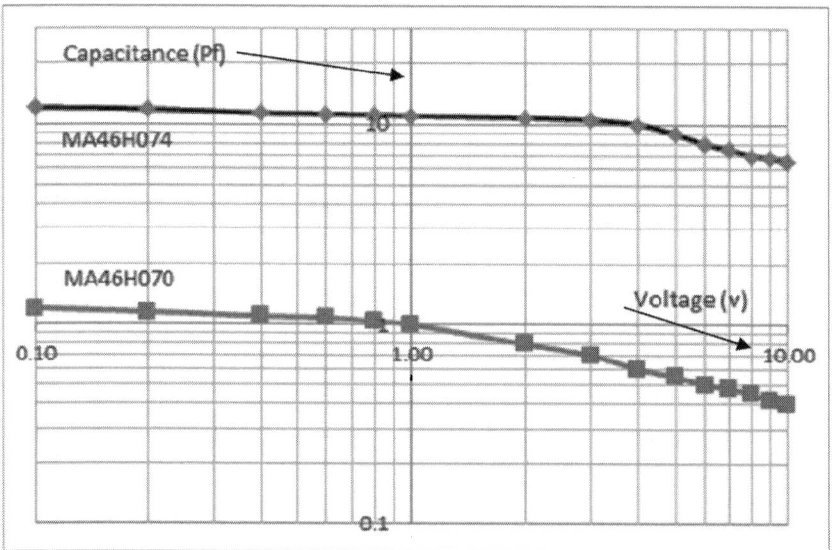

Figure 37. C-V curves of varactors MA46H070 to MA46H076

Figure 38 presents a compact tunable antenna with a varactor. Figure 39. presents measured S_{11} as function of varactor bias voltage. We may conclude that varactors may be used to compensate variations in the antenna resonant frequency at different locations on the human body.

Figure 38. Tunable antenna with a varactor

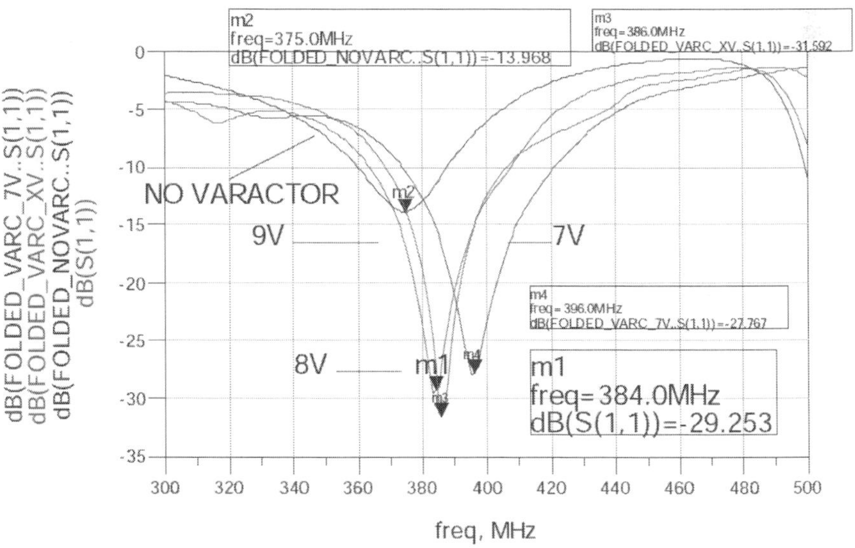

Figure 39. Measured S_{11} as function of varactor bias voltage

8. COMPACT WEARABLE RFID ANTENNAS

RFID (**R**adio **F**requency **Id**entification) is an electronic method of exchanging data over radio frequency waves. There are three major components in RFID system: Transponder (Tag), Antenna and a Controller. The RFID tag, antenna and controller may be assembled on the same board. Microstrip antennas are widely presented in books and papers in the last decade [1-7]. However, compact wearable printed antennas are not widely used at 13.5MHz RIFD systems. HF tags work best at close range but are more effective at penetrating non-metal objects especially objects with high water content.

A new class of wideband compact printed and microstrip antennas for RFID applications is presented in this chapter.

RF transmission properties of human tissues have been investigated in several papers [8-9]. The effect of human body on the antenna performance is investigated in this chapter. The proposed antennas may be used as wearable antennas on persons or animals. The proposed antennas may be attached to cars, trucks, containers and other various objects.

8.1. Dual polarized 13.5MHz compact printed antenna

One of the most critical elements of any RFID system is the electrical performance of its antenna. The antenna is the main component for transferring energy from the transmitter to the passive RFID tags, receiving the transponder's replying signal and avoiding in-band interference from electrical noise and other nearby RFID components. Low profile compact printed antennas are crucial in the development of RIFD systems.

A new compact microstrip loaded dipole antennas has been designed at 13.5MHz to provide horizontal polarization. The antenna consists of two layers. The first layer consists of FR4 0.8mm dielectric substrate. The second layer consists of Kapton 0.8mm dielectric substrate. The substrate thickness determines the antenna bandwidth. A printed slot antenna provides a vertical polarization. The proposed antenna is dual polarized. The printed dipole and the slot antenna provide dual orthogonal polarizations. The dual polarized RFID antenna is shown in Figure 40. The antenna dimensions are 6.4x6.4x0.16cm. The antenna may be attached to the customer shirt in the customer stomach or back zone. The antenna has been analyzed by using Agilent ADS software.

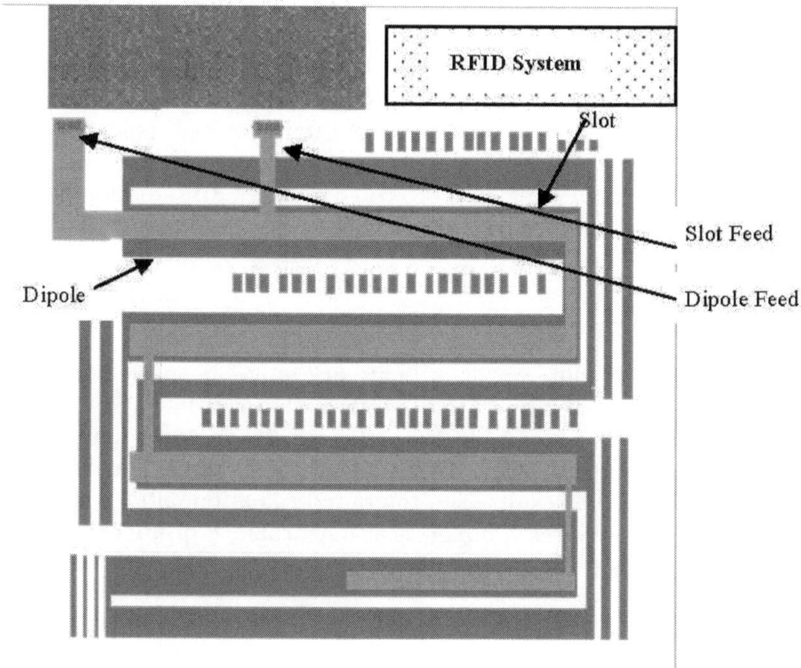

Figure 40. Printed Compact dual polarized antenna, 64x64x1.6mm

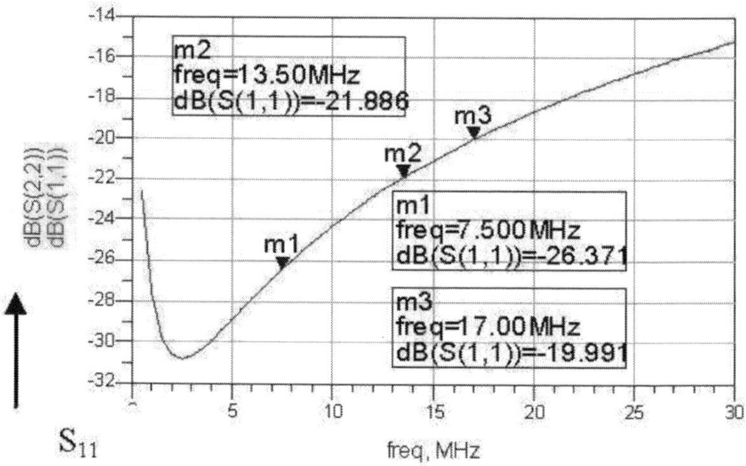

Figure 41. Computed S_{11} results

The antenna S11parameter is better than -21dB at 13.5MHz. The antenna gain is around -10dBi. The antenna beam width is around 160º. The computed S_{11} parameters are presented in Figure 41. There is a good agreement between measured and computed results. Figure 42 presents the antenna measured S_{11} parameters. The antenna cross- polarized field strength may be adjusted by varying the slot feed location. The computed radiation pattern is shown in Figure 43. The computed 3D radiation pattern of the antenna is shown in Figure 44.

8.2. Varying the antenna feed network

Several designs with different feed network have been developed. A compact antenna with different feed network is shown in Figure 45. The antenna dimensions are 8.4x6.4x0.16cm. Figure 46 presents the antenna computed S11on human body. There is a good agreement between measured and computed results. The computed radiation pattern is shown in Fig 47. Table 5 compares the electrical performance of a loop antenna with the compact dual polarized antenna.

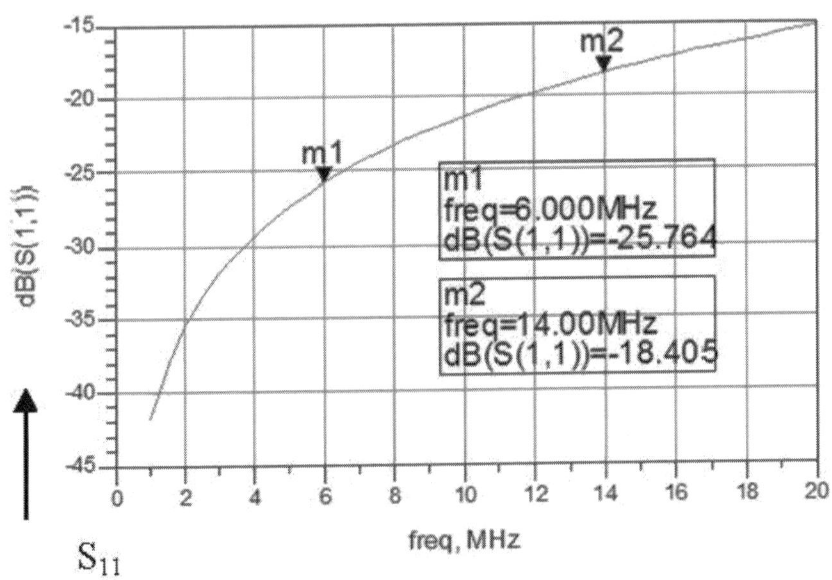

Figure 42. Measured S_{11} on human body

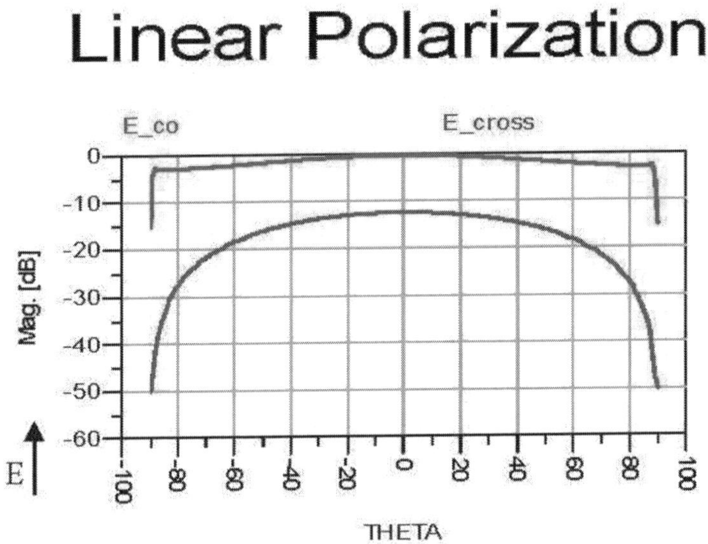

Figure 43. Antenna Radiation pattern

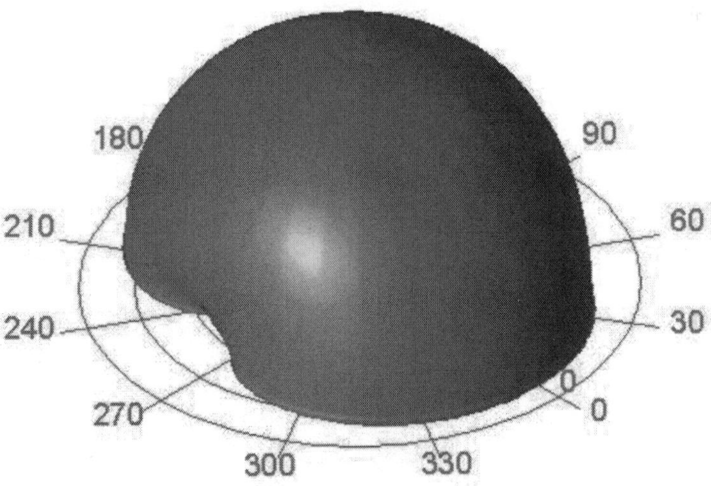

Figure 44. Antenna Radiation pattern

8.3. RFID wearable loop antennas

RFID loop antennas are widely used. Several RFID loop antennas are presented in [15]. RFID loop antennas have low efficiency and narrow bandwidth. As an example the measured impedance of a square four turn loop at 13.5MHz is 0.47 + j107.5 Ω. A matching network is used to match the antenna to 50 Ω. The matching network consists of a 56pF shunt capacitor, 1kΩ shunt resistor and another 56pF capacitor. This matching network has narrow bandwidth.

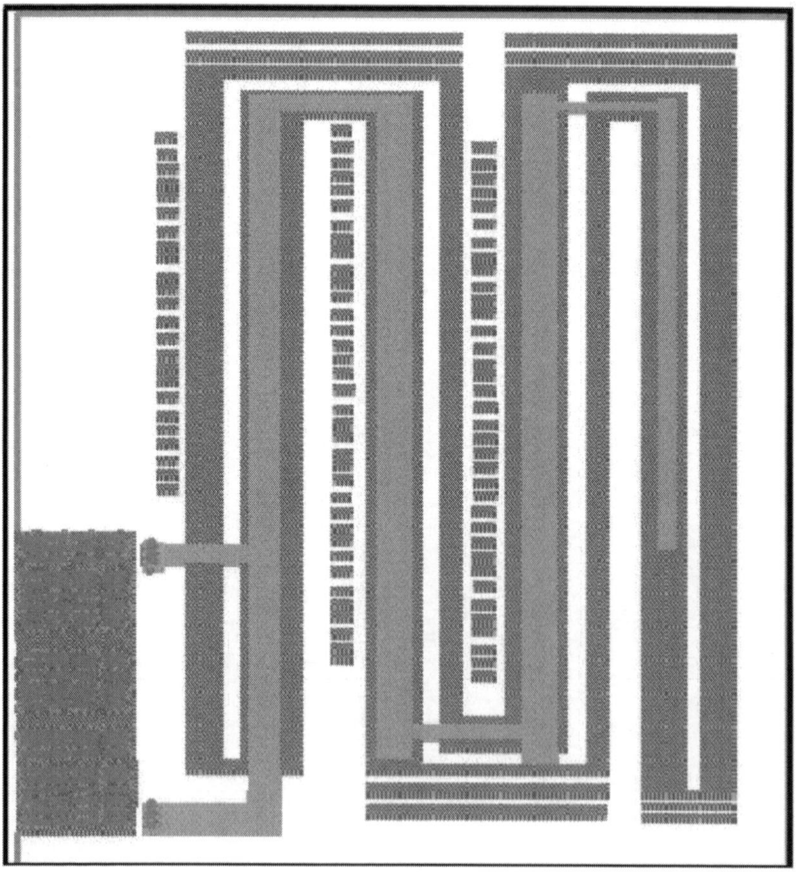

Figure 45. RFID printed antenna, 8.4x6.4x0.16cm.

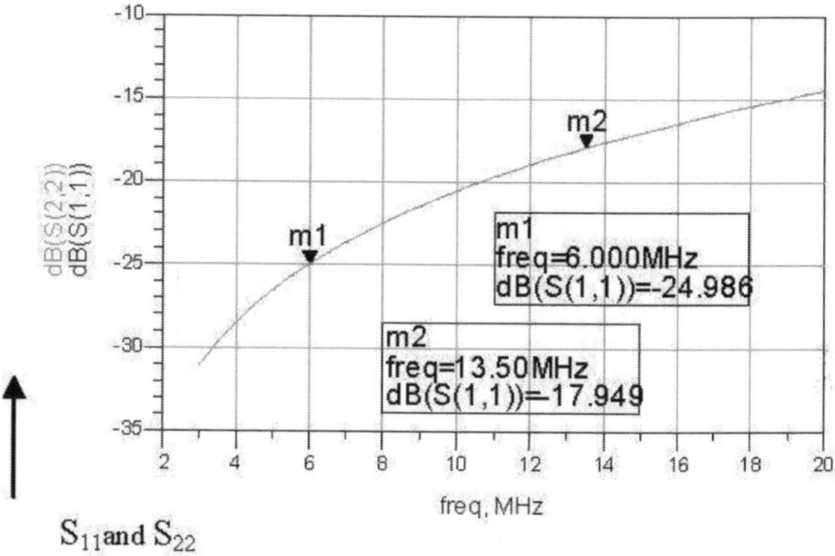

Figure 46. RFID Antenna Computed $S_{11}$and S_{22} results

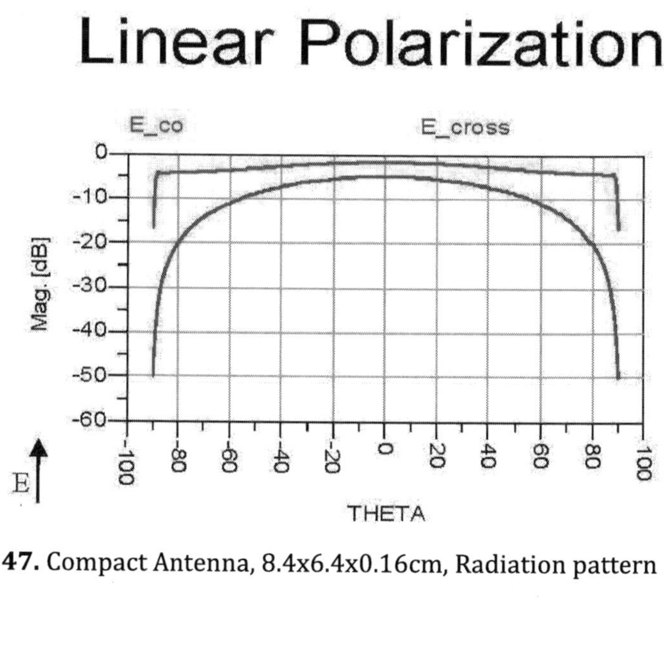

Figure 47. Compact Antenna, 8.4x6.4x0.16cm, Radiation pattern

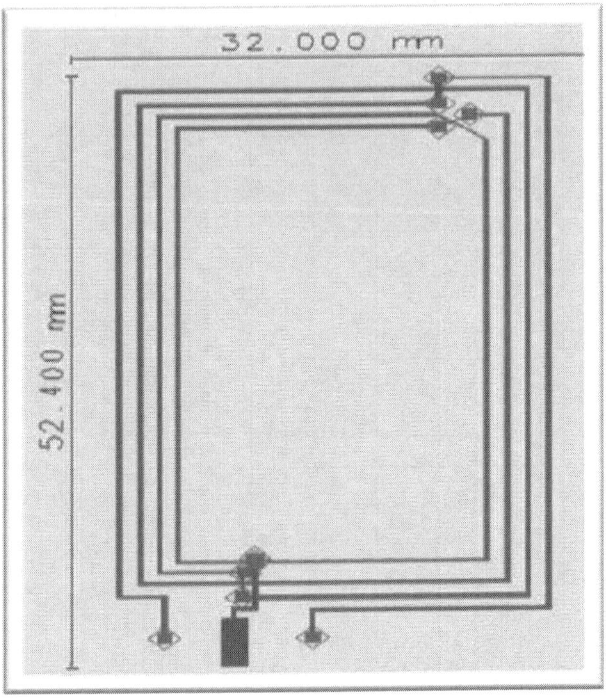

Figure 48. A square four turn loop antenna

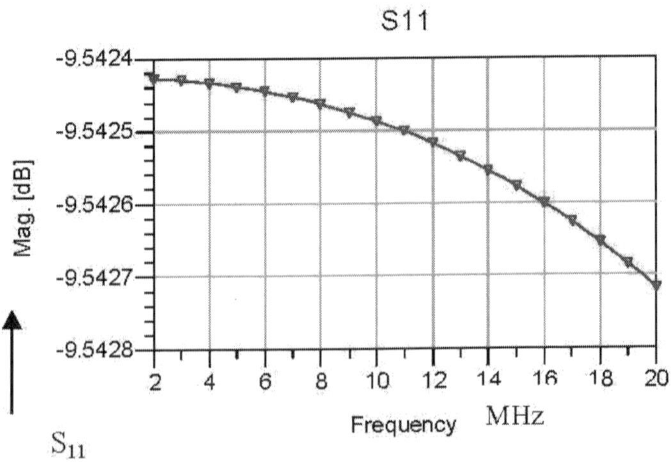

Figure 49. Loop Antenna Computed S_{11} results

A transmitting antenna is placed 20cm from a square four turn loop antenna. 0dBm CW signal is applied to the transmitting antenna. The measured power level at the output of the loop antenna is -50dBm. A square four turn loop antenna has been designed at 13.5MHz by using Agilent ADS software.

The antenna is printed on a FR4 substrate. The antenna dimensions are 32x52.4x0.25mm. The antenna layout is shown in figure 48. S11 results of the printed loop antenna are shown in figure 49. The antenna S_{11} parameter is better than -9.5dB without an external matching network. The computed radiation pattern is shown in Figure 50. The computed radiation pattern takes into account an infinite ground plane.

The Microstrip Antenna input impedance variation as function of distance from the body has been computed by employing ADS software. The analyzed structure is presented in Figure 14 Properties of human body tissues are listed in Table 2 see [8]. These properties were used in the antenna design. S_{11} parameters for different human body thicknesses have been computed. We may note that the differences in the results for body thickness of 15mm to 100mm are negligible. S_{11} parameters for different position relative to the human body have been computed. If the air spacing between the antenna and the human body is increased from 0mm to 10mm the antenna S_{11} parameters may change by less than 1%. The VSWR is better than 1.5:1.

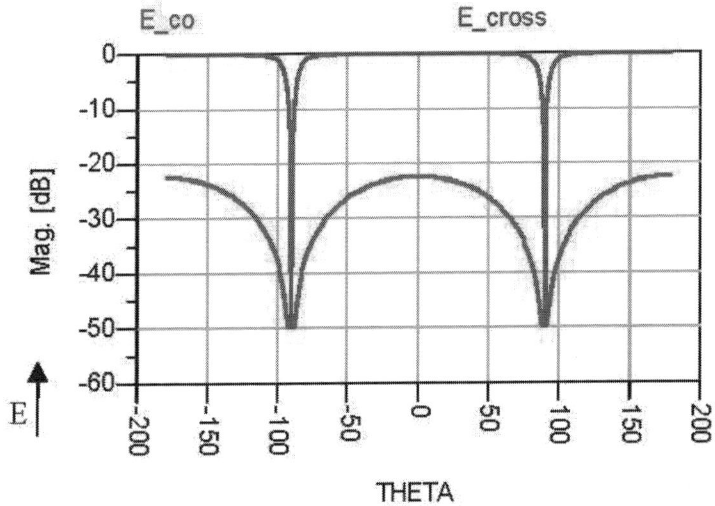

Figure 50. Loop Antenna Radiation patterns for an infinite ground plane

8.4. Proposed antenna applications

An application of the proposed antenna is shown in Figure 51. The RFID antennas may be assembled in a belt and attached to the customer stomach. The antennas may be employed as transmitting or as receiving antennas. The antennas may receive or transmit information to medical systems.

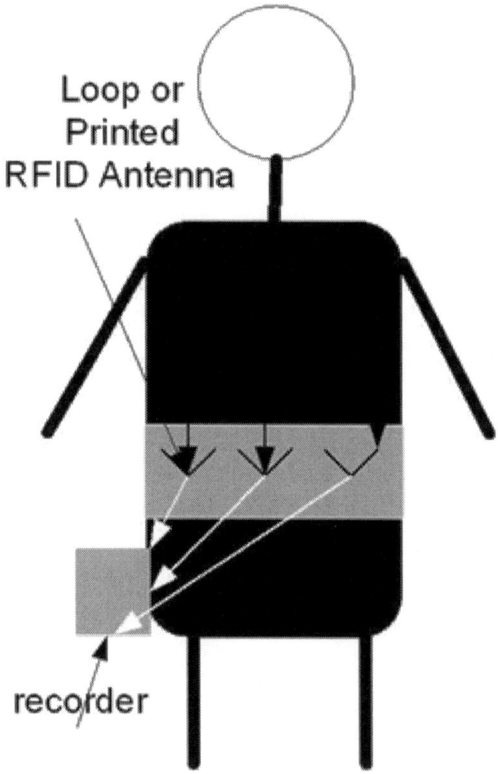

Figure 51. Wearable RFID antenna

In RFID systems the distance between the transmitting and receiving antennas is less than $2D^2/\lambda$, where D is the largest dimension of the antenna. The receiving and transmitting antennas are magnetically coupled. In these applications we refer to the near field and not to the far field radiation pattern.

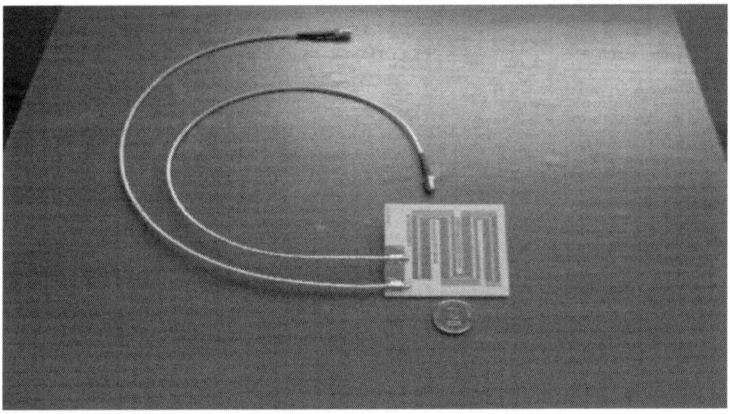

Figure 52. New Microstrip Antenna for RFID applications

Figure 52 and Figure 53 present compact printed antenna for RFID applications. The presented antennas may be assembled in a belt and attached to the patient stomach or back.

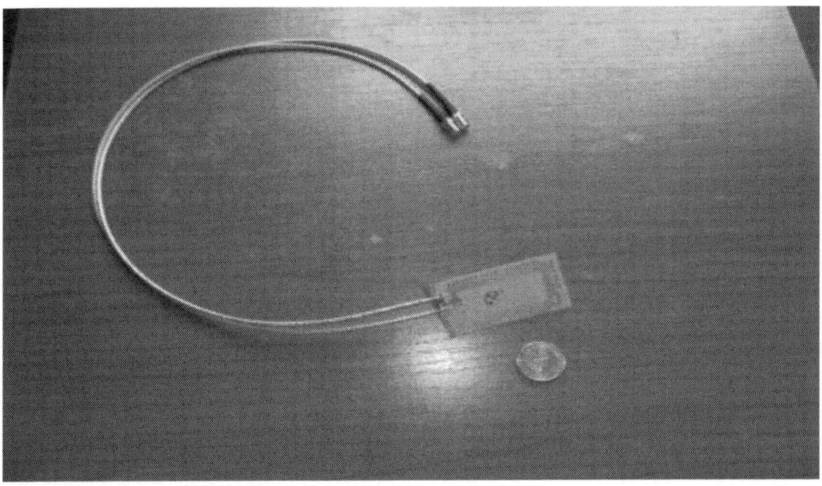

Figure 53. Loop Antenna for RFID applications

9. CONCLUSIONS

This chapter presents wideband microstrip antennas with high efficiency for medical applications. The antenna dimensions may vary from 26x6x0.16cm to 5x5x0.05cm according to the medical system specification. The antennas bandwidth is around 10% for VSWR better than 2:1. The antenna beam width is around 100º. The antennas gain varies from 0 to 4dBi. The antenna S11 results for different belt thickness, shirt thickness and air spacing between the antennas and human body are presented in this chapter. If the air spacing between the new dual polarized antenna and the human body is increased from 0mm to 5mm the antenna resonant frequency is shifted by 5%. However, if the air spacing between the helix antenna and the human body is increased only from 0mm to 2mm the antenna resonant frequency is shifted by 5%. The effect of the antenna location on the human body should be considered in the antenna design process. The proposed antenna may be used in Medicare RF systems.

A wideband tunable microstrip antennas with high efficiency for medical applications has been presented in this chapter. The antenna dimensions may vary from 26x6x0.16cm to 5x5x0.05cm according to the medical system specification. The antennas bandwidth is around 10% for VSWR better than 2:1. The antenna beam width is around 100º. The antennas gain varies from 0 to 2dBi. If the air spacing between the dual polarized antenna and the human body is increased from 0mm to 5mm the antenna resonant frequency is shifted by 5%. A varactor is employed to compensate variations in the antenna resonant frequency at different locations on the human body.

This chapter presents also wideband compact printed antennas, Microstrip and Loop antennas, for RFID applications. The antenna beam width is around 160º. The antenna gain is around -10dBdBi. The proposed antennas may be used as wearable antennas on persons or animals. The proposed antennas may be attached to cars, trucks and other various objects. If the air spacing between the antenna and the human body is increased from 0mm to 10mm the antenna S_{11} parameters may change by less than 1%. The antenna VSWR is better than 1.5:1 for all tested environments.

REFERENCES

1. J.R. James, P.S Hall and C. Wood, "Microstrip Antenna Theory and Design",1981.

2. Sabban and K.C. Gupta, "Characterization of Radiation Loss from Microstrip Discontinuities Using a Multiport Network Modeling Approach", I.E.E.E Trans. on M.T.T, Vol. 39,No. 4,April 1991, pp. 705-712.

3. Sabban," A New Wideband Stacked Microstrip Antenna", I.E.E.E Antenna and Propagation Symp., Houston, Texas, U.S.A, June 1983.

4. Sabban, E. Navon " A MM-Waves Microstrip Antenna Array", I.E.E.E Symposium, Tel-Aviv, March 1983.

5. R. Kastner, E. Heyman, A. Sabban, "Spectral Domain Iterative Analysis of Single and Double-Layered Microstrip Antennas Using the Conjugate Gradient Algorithm", I.E.E.E Trans. on Antennas and Propagation, Vol. 36, No. 9, Sept. 1988, pp. 1204-1212.

6. Sabban, "Wideband Microstrip Antenna Arrays", I.E.E.E Antenna and Propagation Symposium MELCOM, Tel-Aviv,1981.

7. Sabban, "Microstrip Antenna Arrays", Microstrip Antennas, Nasimuddin Nasimuddin (Ed.), ISBN: 978-953-307-247-0, InTech, http://www.intechopen.com/articles/show/title/microstrip-antenna-arrays , pp..361-384, 2011.

8. Lawrence C. Chirwa*, Paul A. Hammond, Scott Roy, and David R. S. Cumming, "Electromagnetic Radiation from Ingested Sources in the Human Intestine between 150 MHz and 1.2 GHz", IEEE Transaction on Biomedical eng., VOL. 50, NO. 4, April 2003, pp 484-492.

9. D.Werber, A. Schwentner, E. M. Biebl, "Investigation of RF transmission properties of human tissues", Adv. Radio Sci., 4, 357–360, 2006.

10. Gupta, B., Sankaralingam S., Dhar, S.,"Development of wearable and implantable antennas in the last decade", Microwave Symposium (MMS), 2010 Mediterranean 2010 , Page(s): 251 - 267.

11. Thalmann T., Popovic Z., Notaros B.M, Mosig, J.R.," Investigation and design of a multi-band wearable antenna", 3rd European Conference on Antennas and Propagation, EuCAP 2009. Pp. 462 - 465.

12. Salonen, P., Rahmat-Samii, Y., Kivikoski, M.," Wearable antennas in the vicinity of human body", IEEE Antennas and Propagation Society International Symposium, 2004. Vol.1 pp. 467 - 470.

13. Kellomaki T., Heikkinen J., Kivikoski, M., " Wearable antennas for FM reception", First European Conference on Antennas and Propagation, EuCAP 2006 , pp. 1-6.

14. Sabban, "Wideband printed antennas for medical applications" APMC 2009 Conference, Singapore, 12/2009.

15. Youbok Lee, "Antenna Circuit Design for RFID Applications", Microchip Technology Inc., Microchip AN 710c.

16. ADS software, Agilent http://www.home.agilent.com/agilent/product.jspx?cc=IL&lc=eng&ckey=1297113&nid=-34346.0. 00&id=1297113

8

CHAPTER

DROOPED MICROSTRIP ANTENNAS FOR GPS MARINE AND AEROSPACE NAVIGATION

Ken G. Clark[1], Hussain M. Al-Rizzo[2], James M. Tranquilla[1], Haider Khaleel[3] and Ayman Abbosh[2]

[1] EMR Microwave Technology Corporation, 64 Alison Blvd., Fredericton, NB, Canada

[2] Systems Engineering Department, Donaghey College of Engineering and Information Technology, University of Arkansas at Little Rock, USA

[3] Department of Engineering Science, Sonoma State University, Rohnert Park, CA, USA

1. INTRODUCTION

The Navigation Satellite Timing and Ranging (NAVSTAR) GPS is a space-based system designed primarily for global real-time, all-weather navigation. There are 30 GPS satellites in six nearly circular, approximately 20,000 kilometer orbital planes, with an inclination of 55^0 relative to the equator [1]. Each satellite transmits two unique, Right Hand Circularly Polarized (RHCP) L band signals. The L1 (1.57542 GHz) carrier is bi-phase modulated with two pseudo-random noise sequences; the P and C/A codes. The L2 (1.2276 GHz) carrier is modulated only with the P code and is used mainly to determine and correct phase advance caused by the ionosphere. Superimposed on the P and C/A codes is the navigation message which contains, among other things, satellite ephemerides, clock biases, and ionosphere correction data [1].

Due to their light weight, reduced size, low cost, conformability, robustness, and ease of integration with MMIC, tremendous research has been reported over the last three decades into the use of microstrip antennas in GPS navigation [2]-[20]. Antenna designers are often faced with interrelated, strict, and conflicting performance requirements in order to meet the accuracy, continuity, and integrity of differential GPS, relative geodetic and hydrographic surveying, ship-borne and aerospace navigation [21]-[27].

The design specifications of a GPS antenna depend on the performance requirements peculiar to the application under consideration. A GPS user antenna requires RHCP and adequate co-polarized radiation pattern coverage over almost the entire upper hemisphere to track all visible satellites. Moreover, the antenna should ideally provide a uniform response in amplitude and, more critically, in phase to the full visible satellite constellation [21]. The angle cutoff and roll-off characteristics of the radiation pattern can be altered to suit the application of interest. For example, fixed ground reference stations and relative static geodetic surveying demand a rapid fall-off near the horizon, a high cross-polarization rejection, and a front-to-back gain ratio in excess of 20 dB to mitigate deleterious effects of severe multipath [28], [29].

In real-time kinematic positioning, few if any of the above constraints may be effective [30] and it may be necessary to operate the antenna under less than optimal conditions in regard to cross polarization performance if a wide beamwidth is of precedence. Precise GPS hydrographic surveying on a vessel cruising at speeds of 10 to 20 knots in open oceans is a challenging task due to the rotational disturbances from a relatively harsh sea environment. Pitch and roll amplitudes as high as 10^0 to 15^0 may be encountered in stormy weather, which presents a major obstacle to GPS derived attitude determination [21], [22]. Another envisaged application for the drooped microstrip antennas introduced in this chapter involves normal pitch or roll maneuvers of a general aviation aircraft, which may cause loss of some satellite signals for a range of flight orientation.

There is significant interest in the commercial and military sectors to develop antennas that could cover much of the upper hemisphere, including GPS satellites at elevation angles as low as 10^0, and to extend the coverage to negative elevation angles [3], [4], [15], [21]. This will lead to fewer occurrences of cycle slips and loss of lock to satellites while rising or setting, will maintain the proper Dilution of Precision (DoP) by maximizing the number of satellites in view, and will reduce the RMS error in range and velocity [1]. Notably, on the negative side, undesired multipath reflections off water and conducting bodies are also strongest at low-elevation angles. Nevertheless, whatever type of antenna is chosen, multipath reception will still have to be dealt with as a common problem [30]. It is fair to say that no single antenna design in the open literature has satisfactorily fulfilled all the above-mentioned requirements on coverage, phase center stability, and multipath rejection for real-time highly dynamic GPS marine and aerospace applications.

A pedestal ground plane is reported in [31], based on a trial-and-error experimental design approach, consisting of a cylindrical structure with a flat elevated center surrounded by sloping sides, to address beam shaping of crossed dipoles. This structure was found to be successful in improving the pattern coverage of the crossed dipole at low-elevation angles. Additional elements were also examined such as folded, serrated, rolled edges, monofilar, and quadrafiliar helices [32]-[35], although none achieved the same degree of radiation pattern control as the crossed dipole. This is attributed in part to the extent of ground plane illumination produced by the different sources and serves to highlight the importance of the ground plane as a secondary source with which to produce pattern changes. Further modifications to the ground plane using choke rings [28] were investigated primarily for multipath rejection.

It is well known that a stand-alone microstrip antenna mounted on a flat ground plane suffers from a lack of pattern control and reduced gain at low-elevation angles. This may result in a loss of contact with satellites when the antenna is mounted on a highly-dynamic platform, an attractive use for this low-profile structure. Fundamental to the design of a patch antenna is the interaction with the ground plane. In fact, the size, orientation, and shape of the ground plane are among the most important parameters that have an influence on the radiation pattern [36]. However, a fundamental distinction exists in the relationship between a patch and the ground plane when compared with helical or dipole elements in that the ground plane of a microstrip antenna forms an integral part of the radiating structure and may not best be defined as a "secondary" source.

Building on our previous design experiences [28], [31]-[35], our research group at the University of New Brunswick, Fredericton, NB, Canada was the first to rigorously investigate the potential performance enhancements and limitations involved when these design modifications are applied to the more appealing microstrip antenna element [37]. The advantages associated with the microstrip antenna are such that one patent has been issued to a GPS manufacturer [38] based only on a downward drooped antenna structure. Neither the dimensions nor the performance of the proposed antennas were quantified in [38]. Later, a corner truncated square patch, partially enclosed within a flatly folded conducting wall, mounted on a pyramidal ground plane, was reported in [39]. However, neither the cross polarization performance nor the phase center stability were provided in [39].

In contrast to the antenna reported in [39], the drooped microstrip antennas introduced in this chapter have the ground plane and actual element deformed such that the corners or edges of the resonant cavity region fall away from the plane occupied by the element. A fundamental understanding of the operation and limitations of the drooped microstrip antenna is still lacking. A diffraction technique was attempted in [40] to model the effects of a sloping ground plane. This, however, was limited by the difficulty of implementing a realistic source term as well as the inclusion of finite lossy dielectric materials. A rigorous full-wave 3-D model, which incorporates the coaxial feed and detailed geometrical features of the drooped microstrip antenna, has not yet been reported.

For these reasons, we have performed the research reported in this chapter, which is the first to our knowledge that combines rigorous 3-D full-wave simulations and experimental measurements to provide a comprehensive characterization of downward and upward drooped microstrip antennas. A FDTD model has been developed, validated experimentally, and used to compute the input impedance and far-field radiation patterns. The FDTD model was used to examine the effects of a wide range of structural variations to gain an insight into the benefits and limitations of the proposed antennas. The parameters of interest include the location and angle of the bend, length of the ground plane, dielectric constant, and thickness of the substrate. Prototype structures were constructed, and their characteristics measured and then compared against simulated results. The authors wish to point out that the antennas described in this chapter are not intended to target multipath mitigation; on the contrary, they demonstrate the range of pattern modifications that could be accomplished by manipulating the orientation and size of the ground plane to suit GPS applications in marine and aerospace navigation.

The rest of the chapter is organized as follows: Section 2 summarizes the FDTD algorithm developed to perform the design and parametric studies and presents results from experimental tests performed to validate the implementation of the model and to demonstrate its ability to correctly predict the behavior of the drooped antennas. Section 3 addresses the design procedure, introduces parametric studies, and describes the drooped antennas constructed and tested for the control of the radiation patterns. Finally, Section 4 provides concluding remarks.

2. THE CONTOUR PATH FDTD MODEL: EXPERIMENTAL VALIDATION

The basic microstrip antenna, which consists of a conducting patch radiator rotated by 45^0 with respect to the center of the ground plane, and separated from the ground plane by a thin dielectric substrate, and the downward drooped antenna are shown in Figure1.

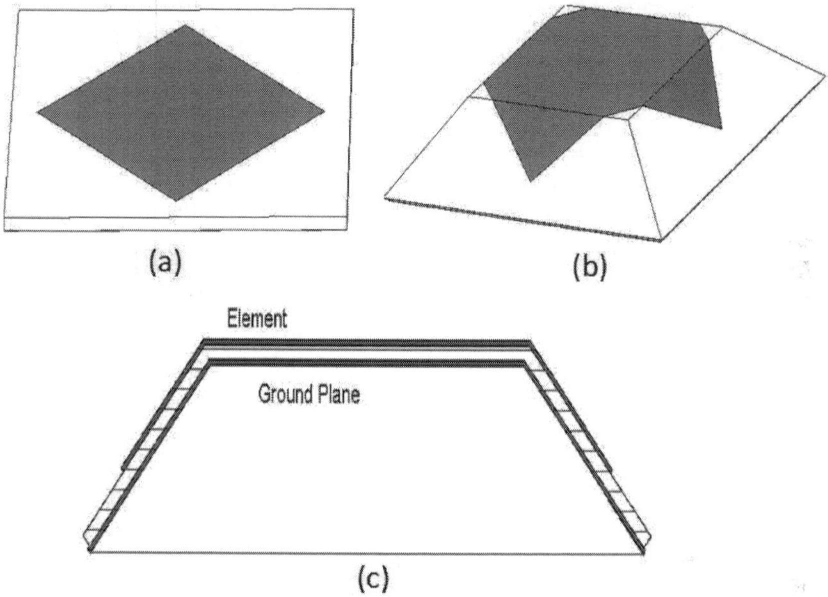

Figure 1. (a) The reference flat microstrip antenna, (b) the downward geometry, (c) cross sectional view of the resonant cavity.

The element is driven by a 50 Ω coaxial cable passing through the ground plane and the substrate. The antenna operates at the L1 GPS frequency of 1.57542 GHz since the majority of commercial GPS receivers use only the L1 frequency. Due to the relatively low cost, time savings, and repeatability, computer simulations can often complement and reduce the empiricism involved in an otherwise purely experimental approach, particularly in the initial design phase, allowing the antennas to be characterized carefully prior to their construction. Because of the complex geometries involved, the task of modeling the drooped microstrip structures is by no means a simple endeavor; it better lends itself to numerical simulation techniques. The FDTD method has been adopted in this research due to its conceptual simplicity and ease of implementation. Because it is a time-domain scheme, it is straightforward to impose a pulse excitation to perform broadband analysis using the Discrete Fourier Transform (DFT).

The FDTD algorithm is implemented in a 3-D Cartesian coordinate system with the formulation allowing for different spatial increments along each coordinate direction. Provision is made for modeling symmetrical objects by applying the Neumann boundary condition along one surface of the computational space. The antenna is excited either by a sinusoidal signal at the resonant frequency of the dominant mode or by a Gaussian pulse with a specified width and delay. The excitation is applied to either the electric or magnetic field, depending on whether a voltage or current source is desired.

In order to compute the input impedance, the instantaneous voltage and current are calculated at a fixed location in the coaxial feed by integrating the radial electric field and the magnetic field components encircling the inner conductor of the coaxial cable. A Gaussian pulse is used, and a DFT is performed to obtain broadband results. The impedance is then translated into the ground plane aperture by standard transmission line methods. The far-field radiation patterns are next determined by driving the model to steady state, using a sinusoidal wave at the fundamental resonant frequency of the antenna.

With pulse excitation, the fields must settle toward zero as energy escapes through the absorbing boundary. The condition used to judge if steady state is reached requires that the energy monitored at several observation points within the computational domain remains below 1% of the peak observed value with lower values enforced for cases continuing to display periodic oscillations in the fields. For a sinusoidal excitation, the solution must converge to an oscillation. The magnitude and phase at several observation points are extracted from the DFT at each temporal cycle. A 1% variation in steady state is permitted in phase.

The methodology we followed involves extracting the antenna characteristics for a given geometry selected from a parametric study using the FDTD code; constructing a prototype; measuring the frequency response of the input impedance, far-field radiation patterns at the measured resonant frequency; and finally comparing simulated results against measurements. To validate the operation of the FDTD model, several antenna structures were simulated and measured. The first antenna used to validate the code was a flat rectangular microstrip. The 50 mm × 47 mm patch was constructed on a square ground plane, 150 mm in side length. The substrate has a relative dielectric constant, ε_r = 4.2 and a thickness of 1.5 mm. The frequency spectrum of the real and imaginary parts of the input impedance and the far-field radiation pattern in the *E* plane are shown in Figs. 2 and 3, respectively.

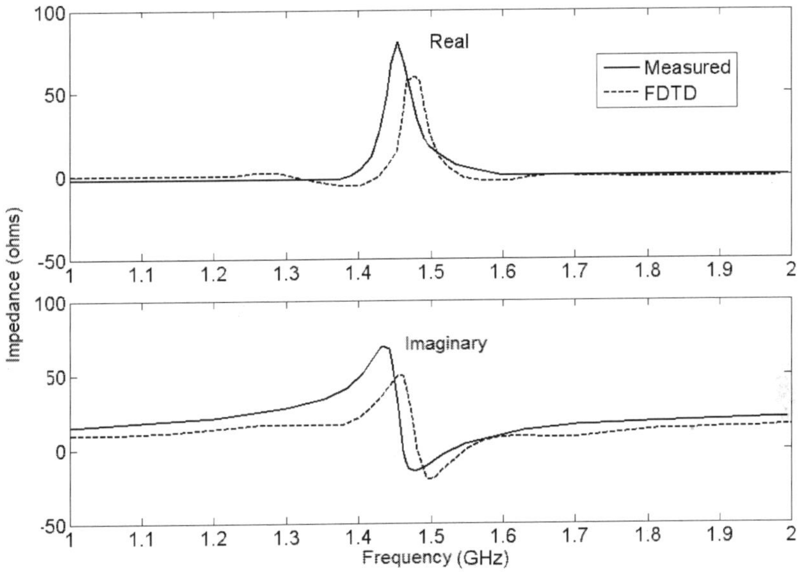

Figure 2. Measured and calculated input impedance for a 50 mm × 47 mm microstrip on a flat, 150 mm square ground plane.

The calculated impedance correctly predicts the resonant frequency measured using a network analyzer. Similarly, excellent agreement is observed between the amplitude and phase of the measured and simulated far-field radiation patterns.

As a progression toward the drooped structure, a microstrip antenna with two edges bent down at an angle of 45^0 was modeled. This structure requires special consideration in that the boundary conditions imposed by the element and ground plane do not fall along the coordinate planes. The most straightforward approach is to discretize the sloping sides with the traditional stair-stepped approximation. However, it has been observed that this can result in a slight change in the resonant frequency of high Q structures [41]. To avoid errors associated with the conventional FDTD method, we used the contour path method introduced in [42]. In this approach, a field component adjacent to a boundary is not updated in terms of the spatial derivatives of the surrounding fields but by an integration of adjacent fields along the perimeter of the cell. This allows for partial or deformed cells, thus better approximating the drooped surfaces.

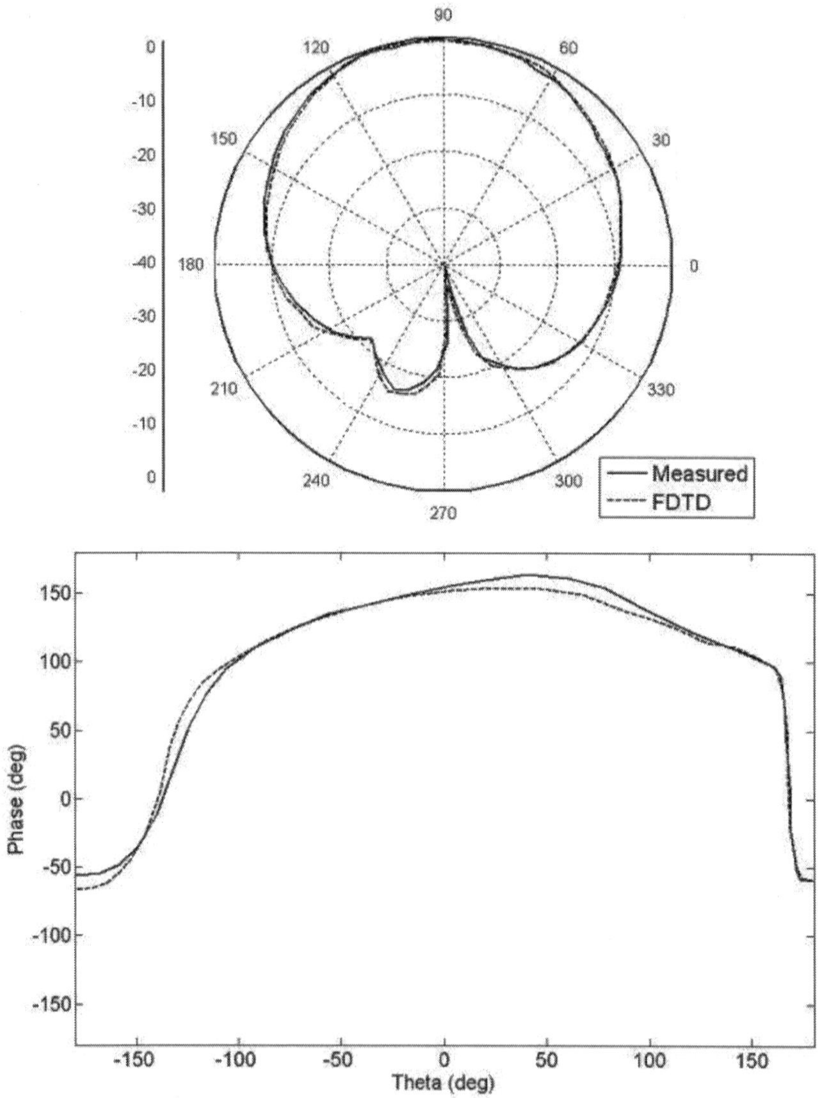

Figure 3. Normalized amplitude and phase of the far-field radiation pattern in the *E* plane for a 50 mm × 47 mm flat microstrip on a 150 mm square ground plane.

Three different step approximations, shown in Figure 4, were used to model the sloping sides, depending upon the angle of the bend. *For fine adjustments*, the ratio of the vertical to horizontal spatial increments is adjusted to yield the desired slope angle. This has an additional benefit of simplifying the implementation of the contour method when applied to field components adjacent to the sloped surfaces. Because each cell is truncated in the same way, the contour corrections proceeded without the need to calculate intercepts of the actual slope line at each cell.

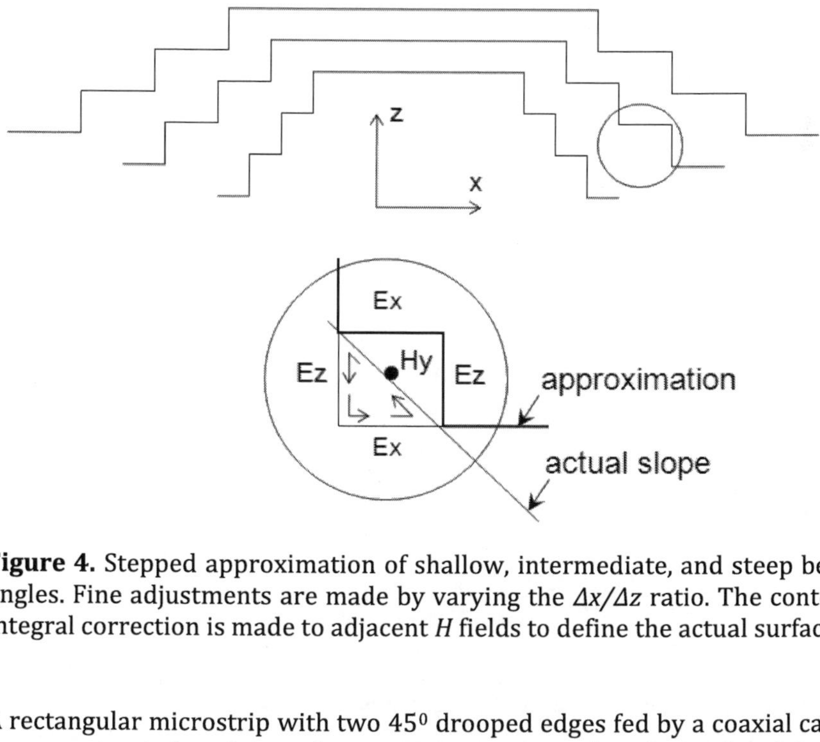

Figure 4. Stepped approximation of shallow, intermediate, and steep bend angles. Fine adjustments are made by varying the $\Delta x/\Delta z$ ratio. The contour integral correction is made to adjacent *H* fields to define the actual surface.

A rectangular microstrip with two 45^0 drooped edges fed by a coaxial cable was constructed and tested to compare measured and simulated input impedance, *E* and *H* plane patterns. The geometry of the antenna is shown in Figure 5 along with the phase of the *E* plane elevation cut. The measured phase displayed in Figure 5 showed a slight asymmetry due to the offset in antenna mount necessary to accommodate the bends and the connector. For comparison, we referenced the calculated far-field patterns to the same offset origin. The *E* plane and *H* plane patterns shown in Figs. 5 and 6 along with the impedance of Figure 7 reveal good correspondences between measured and simulated results.

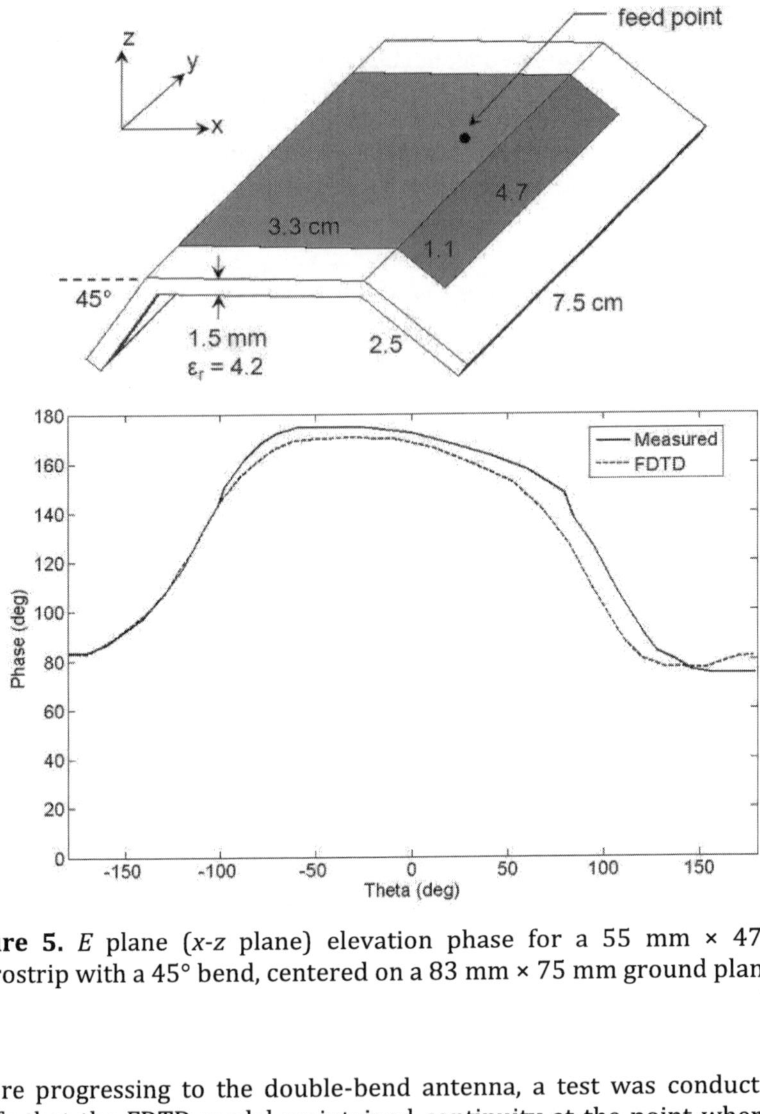

Figure 5. *E* plane (*x-z* plane) elevation phase for a 55 mm × 47 mm microstrip with a 45° bend, centered on a 83 mm × 75 mm ground plane.

Before progressing to the double-bend antenna, a test was conducted to verify that the FDTD model maintained continuity at the point where the stepped approximation changed from a 2:1 to a 1:1 ratio. To accomplish this, we modeled a 40 mm × 50 mm bent microstrip antenna near the change of an angle of 55^0 using both ratios. A comparison of the input impedance and elevation patterns depicted in Figure 8 show good agreement between results obtained from each approximation.

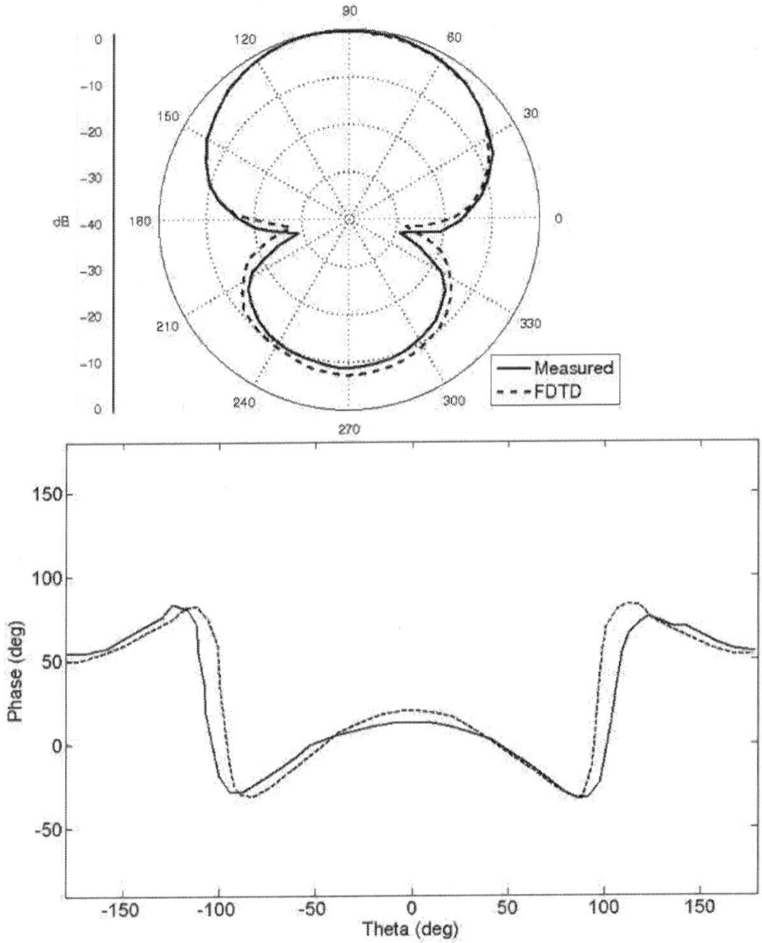

Figure 6. *H* plane elevation pattern for a 55 mm × 47 mm microstrip with a 45° bend, 10 mm from each short side, magnitude (top) and phase (bottom).

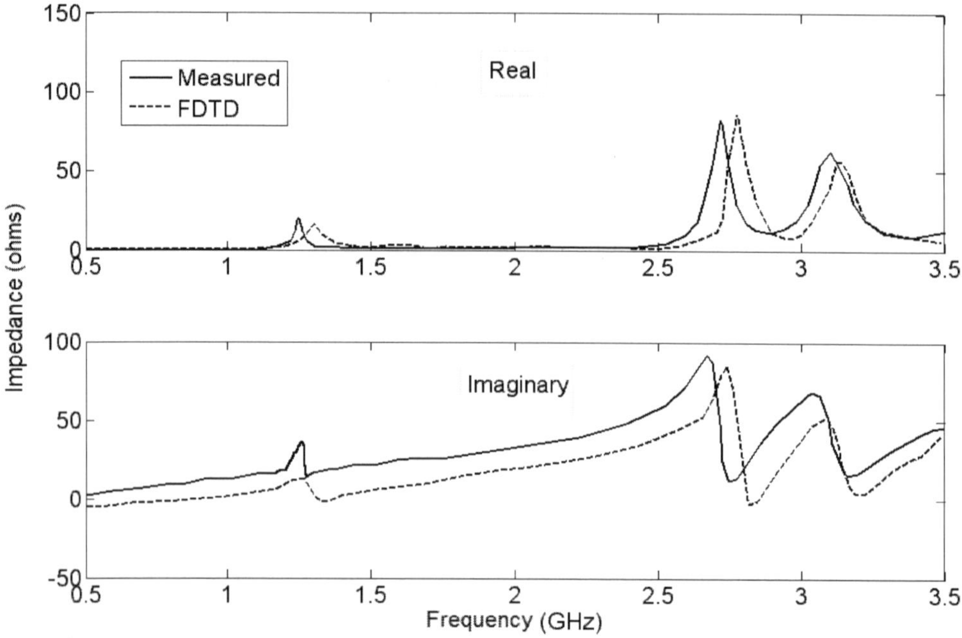

Figure 7. Measured and calculated input impedance for a 47 mm × 55 mm microstrip on a 45° bent ground plane (33 mm flat top with 11 mm bent over each angle).

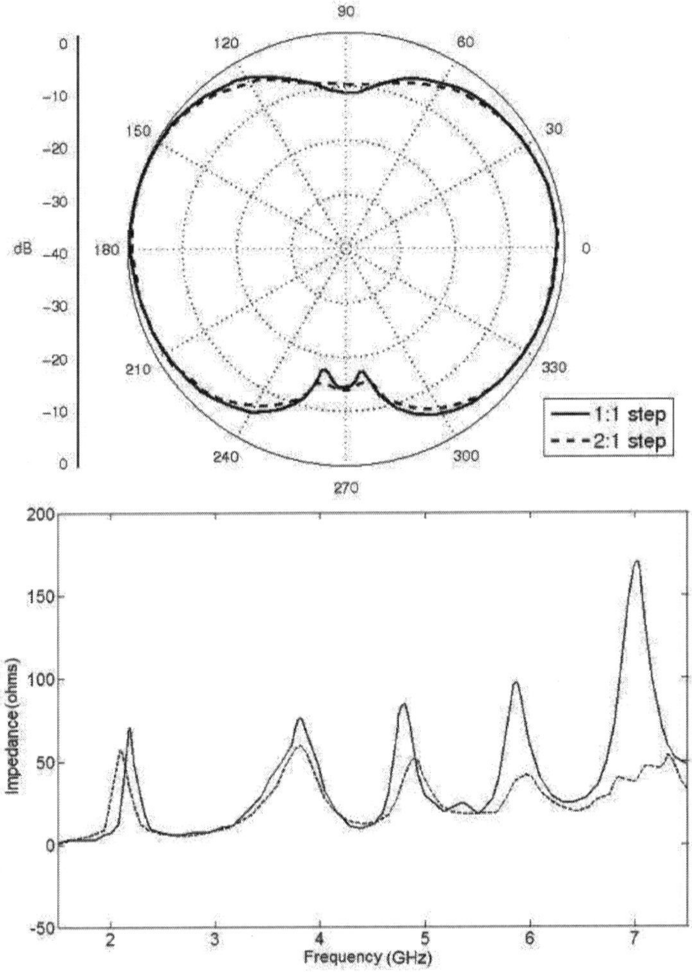

Figure 8. *E* plane pattern and impedance comparison for 50 mm × 40 mm microstrip (ε_r = 2) with a 55° bend, calculated for a 1:1 and 2:1 step approximation of the sloped sides.

To duplicate the bend on the two remaining sides to achieve the full drooped structure, we constructed two drooped antennas and used them to verify the performance of the completed model. The first antenna, shown in Figure 9, consists of a 62 mm × 62 mm patch, a 40 mm square elevated section, printed on a 1.5 mm thick substrate, ε_r = 4.2, and a 60^0 droop angle. The second is a 64 mm × 64 mm patch, a 50 mm square elevated section, printed on a 3 mm substrate, ε_r = 2.2, and a 30⁰ droop angle. Results obtained for the real and imaginary parts of the input impedance and far-field radiations patterns, as shown in Figs. 9 and 10, respectively, display good agreement between simulated and measured results. The excellent agreement demonstrated thus far between the amplitude and phase of the simulated and measured far-field radiation patterns prompted further exploration of the possibility of controlling the radiation pattern by manipulating the droop parameters.

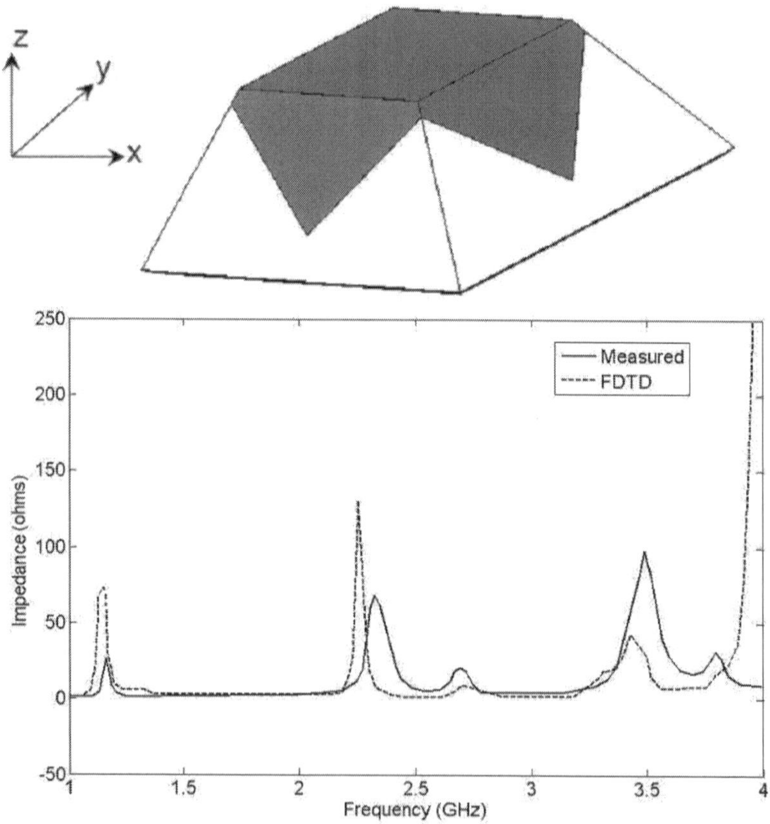

Figure 9. Measured and calculated input impedance for a 60° double bend microstrip.

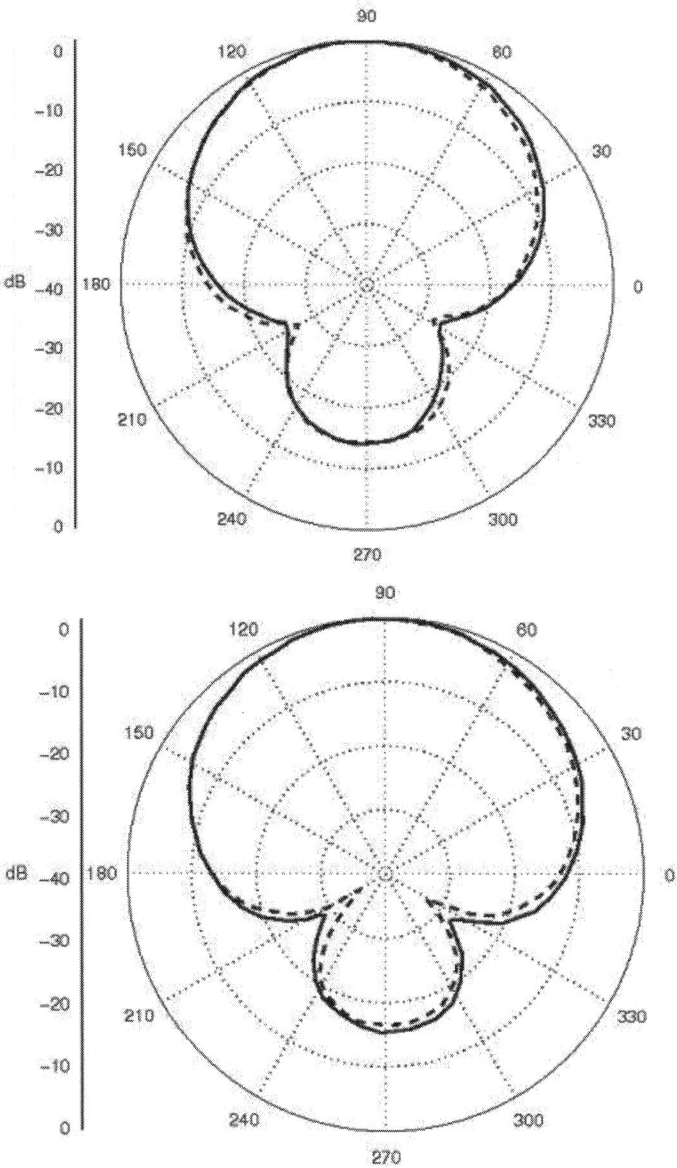

Figure 10. Measured and calculated elevation patterns for 64 mm square microstrip with a 30^0 bend, ε_r = 2.2, *E* plane (*x-z*) top and *H* plane (*y-z*) bottom (1550 MHz).

3. PARAMETRIC ANALYSIS OF THE DROOPED MICROSTRIP ANTENNAS

Further parametric analyses were conducted to determine the range of structural variations that can be utilized to optimize the performance of the drooped microstrip antennas. The design process was complicated due to the several interacting parameters that must be considered in order to provide adequate low-angle coverage, uniform phase response, and polarization purity. Parameters investigated include the angle and location of the bend, length of the ground plane skirt, thickness and dielectric constant of the substrate. The results are assessed on the basis of their impact on antenna gain at bore sight, phase performance in the upper hemisphere, pattern beamwidth, cross polarization rejection, and near horizon gain reduction.

In all cases to be presented, the radiation patterns were obtained at the resonant frequency of the dominant mode. The amplitude and phase of the radiation patterns were obtained from anechoic chamber measurements in a $5^0 \times 5^0$ grid, following the procedure described in [43]. These were then analyzed to determine the 3 dB beamwidth and near horizon gain reduction with respect to bore sight (zenith). In order to determine the absolute gain, we integrated the calculated patterns over the upper and lower hemispheres with a 5^0 step in azimuth and elevation. All field calculations were then referenced to the resulting isotropic power density.

Whereas an enormous volume of literature is available on patch antennas for GPS applications [2]-[20], a close scrutiny revealed that the design objectives in the majority of these studies and the performance characterization were based entirely on the amplitude of the co- and cross-polarized radiation patterns. Only few have considered the phase response as a figure of merit in the design process and/or in the analysis or measurements [28], [31]-[35], [43]. The phase response directly weighs the arriving signals and produces a phase-shaping effect, which depends on the angle of arrival of the satellite signals. The calibration of the phase response provides invaluable information regarding the level of accuracy one can ultimately achieve for sub-centimeter static geodetic positions [43].

The measured upper hemispherical phase response of the antennas under test was matched to an ideal hemisphere in a $5^0 \times 5^0$ grid, using equal solid angle weighting. The position of the ideal hemisphere was adjusted to minimize the RMS error between the measured and ideal phase [43]. The origin of this hemisphere is defined as the "center of best fit" or "phase center," and the difference between the measured and ideal phase is defined as the "phase residual" or "phase error." The RMS value of the phase error is used as a figure of merit to describe the phase distortion introduced by the antennas considered in the rest of this chapter.

3.1. Drooped microstrip with a downward bend

For the initial test, a 40 mm × 40 mm patch was placed on three different ground planes. These, in turn, were bent at three different distances from the center, forming a flat square top that was 10, 30, or 50 mm^2. Simulations were carried out for bend angles, ranging from 0 to 90^0 in a 15^0 step. Figure 11 depicts the parameters and geometries used for the initial set of simulations.

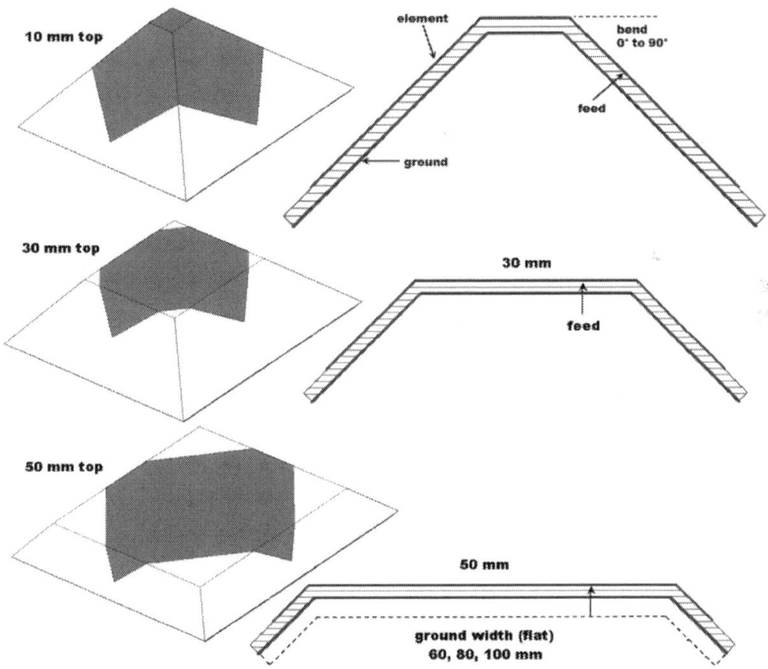

Figure 11. Initial structural variations included three bend locations at 5, 10, and 25 mm from the patch center, droop angles ranging from 0° to 90° and 3 ground planes having flat (unbent) dimensions of 60, 80, 100 mm. Three substrate materials with relative permittivity of 2.2, 4.2, and 10.

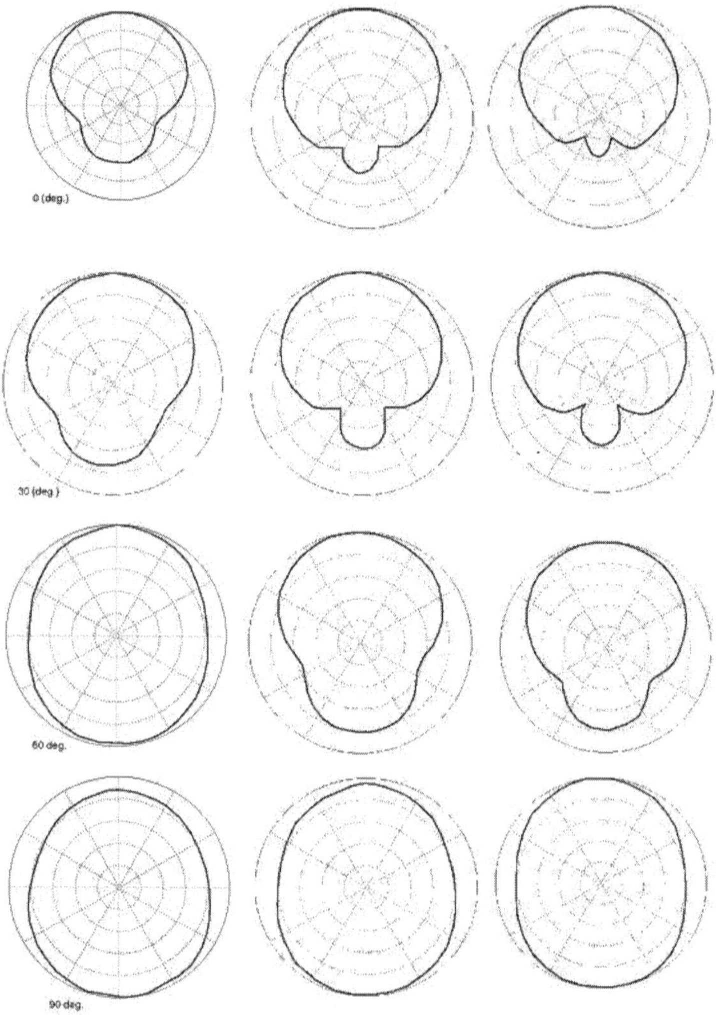

Figure 12. Elevation patterns for bend angle variations of 0° to 90° with 60 (left), 80 (center), and 100 (right) mm ground planes. E_θ component, ε_r = 2.2, all with 30 mm top. All patterns are normalized to 0 dB maximum, 10 dB/division.

Three dielectric constants were used in the course of the simulations to examine the effects of different substrate materials; values of 2.2 and 4.2 were selected because of the availability of substrates to construct the verification cases, while an ε_r of 10 was chosen as an example of a ceramic substrate frequently used in industry. Copper tape applied to bulk Teflon, one eighth inch thick, was used to form several fixed and adjustable structures. A standard polyester-based circuit board was used to construct others.

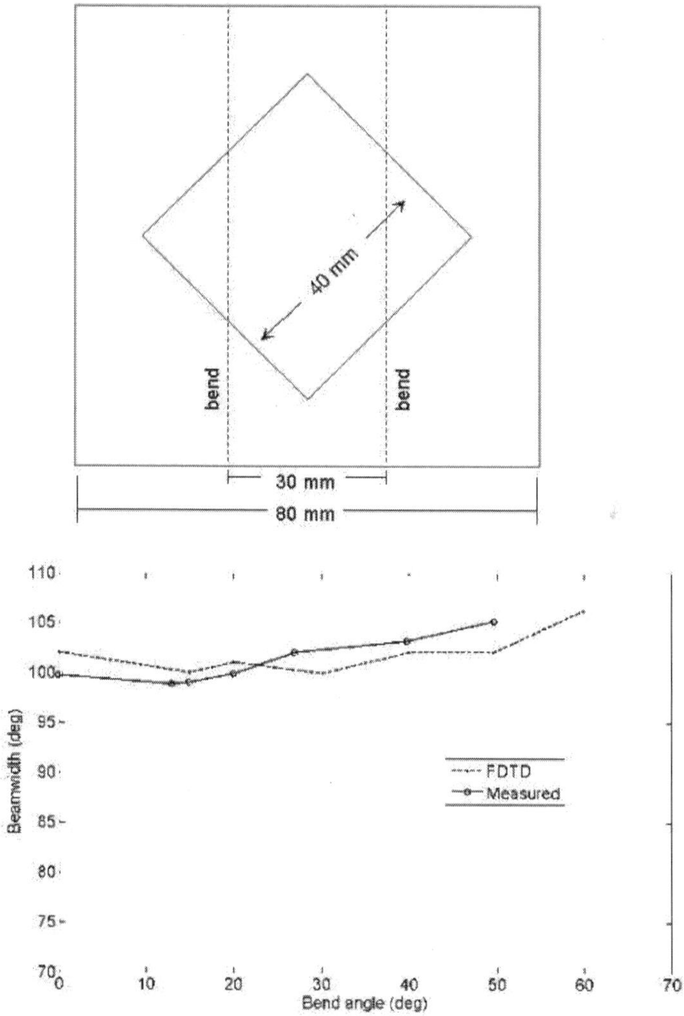

Figure 13. Measured and computed 3 dB beamwidth for an adjustable bend microstrip antenna.

A sequence of elevation patterns is presented in Figure 12, which displays the progression of the radiated fields with the size and bend angle of the ground plane. Not unexpectedly, smaller ground planes with larger bends allow more energy to escape off the back, until it appears that the main beam is 180^0 from the bore sight direction. This is made obvious by plotting the polar patterns for E_θ, using a linearly polarized excitation. With a circularly polarized excitation, backward radiated energy appears predominately in the cross polarized component, making the effect less noticeable. Notably, all subsequent results for gain, beamwidth, or phase are presented while exciting circular polarization as this would be the normal operating mode of the antenna.

Table 1. esults for the drooped microstrip, substrate permittivity 2.2.

ε_r=2.2		Boresight gain (dBi)			3 dB Beamwidth (deg)			Near horizon gain Roll-off (dB)			RMS phase error (deg)		
Width-mm		60	80	100	60	80	100	60	80	100	60	80	100
Bend angle (10 mm top)	0	7.2	7.5	7.7	88	83	80	-11.	-12.	-13.	3.0	2.6	2.0
	15	7.1	7.5	7.7	88	84	80	-10.	-11.	-13.	2.5	3.3	3.4
	30	6.9	7.1	7.1	90	86	86	-9.7	-10.	-11.	1.5	1.8	1.7
	45	5.5	6.5	6.4	96	95	96	-9.0	-9.1	-9.1	2.2	1.6	2.0
	60	0.9	5.6	6.2	100	94	97	-7.9	-9.5	-9.0	4.3	6.5	2.6
	70	0.8	3.3	5.4	119	91	95	-5.5	-9.3	-9.0	12.	4.4	2.8
	90	-24	-25	-26	354	356	356	22	23	24	11.9	2.7	2.8
ε_r=2.2		Boresight gain (dBi)			3 dB Beamwidth (deg)			Near horizon gain Roll-off (dB)			RMS phase error (deg)		
Width-mm		60	80	100	60	80	100	60	80	100	60	80	100
Bend angle (30 mm top)	0	7.2	7.5	7.7	88	83	80	-11.	-12.	-13.	2.9	2.6	2.0
	15	7.1	7.5	7.8	88	83	78	-11.	-11.	-13.	2.4	3.3	3.4
	30	6.9	7.4	7.4	89	83	82	-9.6	-11.	-11.	1.5	1.8	1.7
	45	6.6	7.1	7.0	90	87	87	-10.	-10.	-11.	1.9	1.6	2.0
	60	5.5	6.7	6.5	91	90	93	-10.	-10.	-9.7	2.3	6.5	2.6
	70	5.2	6.4	6.2	92	92	96	-10.	-9.9	-9.1	1.6	4.4	2.8
	90	0.1	2.3	3.78	97	84	85	-9.6	-11	-12	8.2	2.7	2.8
Bend angle (50 mm top)	0	7.2	7.5	7.7	88	83	80	-11.	-12.	-13.	2.9	2.6	2.0
	15	7.0	7.7	7.9	89	81	77	-10.	-12.	-13.	2.4	3.3	3.4
	30	7.1	7.5	7.4	87	82	83	-9.9	-11.	-11.	1.5	1.8	1.7
	45	6.9	7.2	7.1	89	86	87	-9.6	-10.	-10.	1.9	1.6	2.0
	60	6.8	7.2	7.0	89	87	88	-9.7	-10.	-10.	2.3	6.5	2.6
	70	7.1	7.2	6.8	87	87	91	-10.	-10.	-9.4	1.6	4.4	2.8
	90	6.4	6.9	6.9	91	89	89	-9.4	-9.7	-10.	8.2	2.7	2.8

Table 2. Results for the drooped microstrip, substrate permittivity 4.2.

ε_r=4.2		Boresight gain (dBi)			3 dB Beamwidth (deg)			Near horizon gain Roll-off (dB)			RMS phase error (deg)		
Width-mm		60	80	100	60	80	100	60	80	100	60	80	100
Bend angle (10 mm top)	0	4.7	5.8	6.2	106	100	100	-11.6	-13.5	-14.5	1.6	1.2	1.1
	15	4.5	5.7	6.0	106	103	100	-9.9	-10.8	-10.9	2.2	3.6	4.3
	30	4.8	5.9	6.0	106	100	101	-8.6	-11.0	-10.7	1.5	1.0	1.0
	45	3.4	5.2	6.5	149	107	108	-3.7	-9.0	-9.7	2.0	1.5	1.2
	60	0.9	4.3	6.1	360	129	115	2.5	-5.7	-8.4	3.6	2.7	2.6
	70	-1.1	1.5	3.3	360	360	144	6.5	0.8	-3.8	12.8	2.9	3.3
	90	-27	-28	-29	360	360	360	-	-	-	-	-	-
Bend angle (30 mm top)	0	4.7	5.8	6.2	114	106	100	-11.6	-13.5	-14.5	1.6	1.2	1.1
	15	4.5	6.0	6.0	119	103	103	-10.6	-13.8	-13.7	1.2	1.1	0.9
	30	4.6	5.9	6.0	113	104	102	-11.1	-13.3	-13.5	0.9	0.5	0.5
	45	4.9	5.8	5.9	107	104	106	-12.2	-13.4	-13.4	1.2	0.8	0.7
	60	3.5	4.4	5.3	106	109	112	-11.3	-11.4	-12.0	1.6	1.9	1.3
	70	3.0	4.3	4.8	116	114	115	-9.2	-10.9	-11.3	0.9	0.9	0.8
	90	1.0	1.5	2.4	121	110	103	-6.7	-8.2	-10.1	1.6	1.5	1.4

ε_r=4.2		Boresight gain (dBi)			3 dB Beamwidth (deg)			Near horizon gain Roll-off (dB)			RMS phase error (deg)		
Width-mm		60	80	100	60	80	100	60	80	100	60	80	100
Bend angle (50 mm top)	0	4.8	5.8	6.2	114	106	100	-11.6	-13.5	-14.5	0.8	0.9	1.2
	15	4.9	5.9	6.3	119	103	103	-11.5	-13.6	-14.6	0.7	0.7	0.9
	30	5.3	6.0	6.2	113	104	102	-12.5	-13.6	-14.1	0.6	0.5	0.6
	45	5.0	5.8	6.0	107	104	106	-12.4	-13.4	-13.4	0.7	0.6	0.6
	60	4.7	5.6	5.8	106	109	112	-11.7	-12.8	-13.3	0.6	0.6	1.0
	70	5.2	5.5	5.4	116	114	115	-12.2	-12.5	-12.0	0.4	0.4	0.5
	90	4.4	5.2	5.4	121	110	103	-11.6	-12.6	-12.8	0.8	0.6	0.6

Next, a prototype was constructed with adjustable bend plates, and the measured beamwidths were compared against simulated results. For construction simplicity, the antenna was excited using linear polarization with only two sloped sides. In Figure 13, the measured and computed beamwidths show a slight improvement with increasing bend angle.

Variations in bore sight gain, 3 dB beamwidth, near-horizon gain roll-off, and RMS phase error as a result of increasing the bend angle are summarized in Tables 1 to 3. The 10 mm case shows a sharp gain reduction at extreme bend angles. The results reveal a relatively minor improvement in the pattern beamwidth, even for large bend angles and varying ground plane sizes for the higher dielectric substrates. Indeed, the radiation patterns above the horizon remain virtually unchanged for bend angles up to 60^0.

Table 3. Results for the drooped microstrip, substrate permittivity 10.

ε_r=10		Boresight gain (dBi)			3 dB Beamwidth (deg)			Near horizon gain Roll-off (dB)			RMS phase error (deg)		
	Width-mm	60	80	100	60	80	100	60	80	100	60	80	100
Bend angle (10 mm top)	0	3.1	4.7	5.3	114	108	104	-9.4	-11.5	-12.5	0.80	0.49	0.53
	15	3.4	4.4	5.2	116	115	111	-9.3	-10.4	-11.4	0.89	1.0	0.87
	30	3.8	4.4	5.1	117	113	111	-9.7	-10.7	-11.3	0.59	0.46	0.42
	45	2.5	3.6	4.5	123	117	116	-7.8	-9.5	-10.6	0.41	0.41	0.38
	60	3.3	2.6	3.6	139	120	118	-8.2	-8.3	-9.5	3.0	0.80	0.86
	70	2.4	3.0	3.5	134	122	120	-7.4	-8.5	-9.3	2.0	0.72	0.97
	90	-38	-40	-43	360	360	360	-	-	-	-	-	-
Bend angle (30 mm top)	0	3.1	4.7	5.3	114	108	104	-9.4	-11.5	-12.5	0.80	0.49	0.53
	15	3.8	4.5	4.9	119	108	106	-9.6	-11.3	-11.9	0.40	0.46	0.48
	30	3.8	5.0	5.4	116	106	106	-9.8	-12.1	-12.5	0.25	0.33	0.36
	45	2.9	4.1	4.7	116	114	114	-8.9	-10.3	-10.9	0.39	0.32	0.30
	60	2.7	3.6	4.3	119	104	104	-8.6	-11.1	-12.0	0.56	1.2	1.2
	70	2.3	2.9	3.5	122	119	119	-7.8	-8.6	-9.3	0.24	0.21	0.18
	90	1.5	1.4	1.5	134	132	127	-6.0	-6.1	-6.6	0.29	0.31	0.34
Bend angle (50 mm top)	0	3.1	4.7	5.3	114	108	104	-9.4	-11.5	-12.5	0.80	0.49	.53
	15	2.9	4.6	5.4	133	114	109	-7.6	-10.7	-11.7	0.34	0.70	0.80
	30	3.2	4.9	5.5	121	110	109	-8.7	-12.4	-11.8	0.46	0.90	0.32
	45	3.4	4.5	4.8	119	112	117	-9.0	-10.7	-10.4	0.37	0.56	0.41
	60	3.0	3.7	4.9	122	124	119	-8.4	-8.9	-11.8	0.51	0.51	0.90
	70	3.7	3.6	4.5	117	123	107	-9.5	-9.1	-10.3	0.28	2.2	1.5
	90	2.0	2.6	3.2	113	110	108	-8.3	-9.2	-10.0	2.1	2.4	2.2

The phase of the elevation cuts shown in Figure 14 shows a remarkable change in the below-horizon phase. For small bend angles, the phase diminishes from the bore sight value when approaching the horizon, while at higher bends the phase increases from the bore sight value. At about 15^0 to 30^0 of bend, a region exists where the elevation phase remains relatively constant above the horizon. Bending of the structure in this manner could provide an additional beamwidth and, more importantly, the phase stability necessary to achieve a design specification, particularly if further modification of the substrate is not feasible.

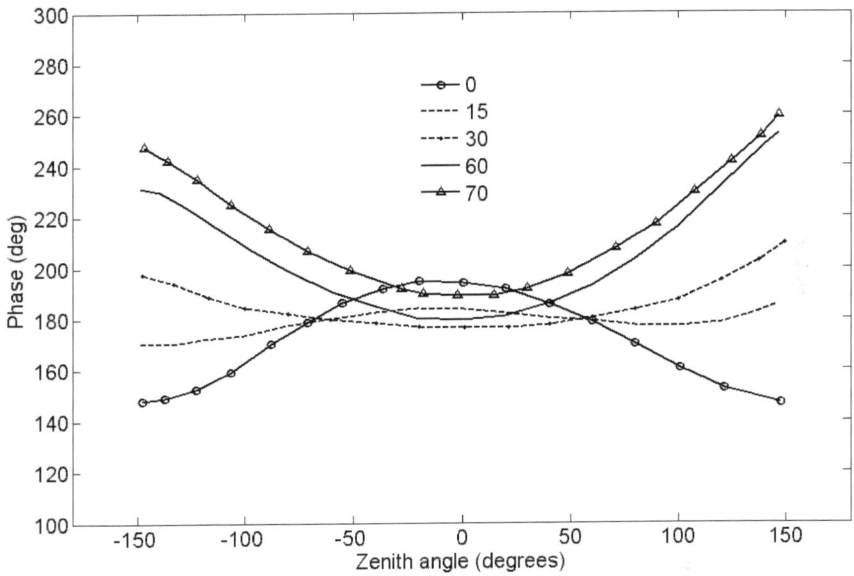

Figure 14. Elevation phase patterns for different bend angles (0⁰, 15⁰, 30⁰, 60⁰, and 70⁰), with a 100 mm ground plane and a 30 mm flat ($\varepsilon_r = 4.2$).

Another consideration for GPS antennas is the cross polarization behavior. Odd reflections from nearby objects tend to be orthogonally polarized. Hence, it is important that the antenna be able to reject these, particularly near the horizon. To demonstrate the effect of bending the structure on the polarization performance, we examined the ratio of the right- to the left-hand circular components for several bend angles. Almost without exception, the bend degraded the cross polarization rejection near the horizon as shown in Figure 15, making the antenna more susceptible to spurious signals.

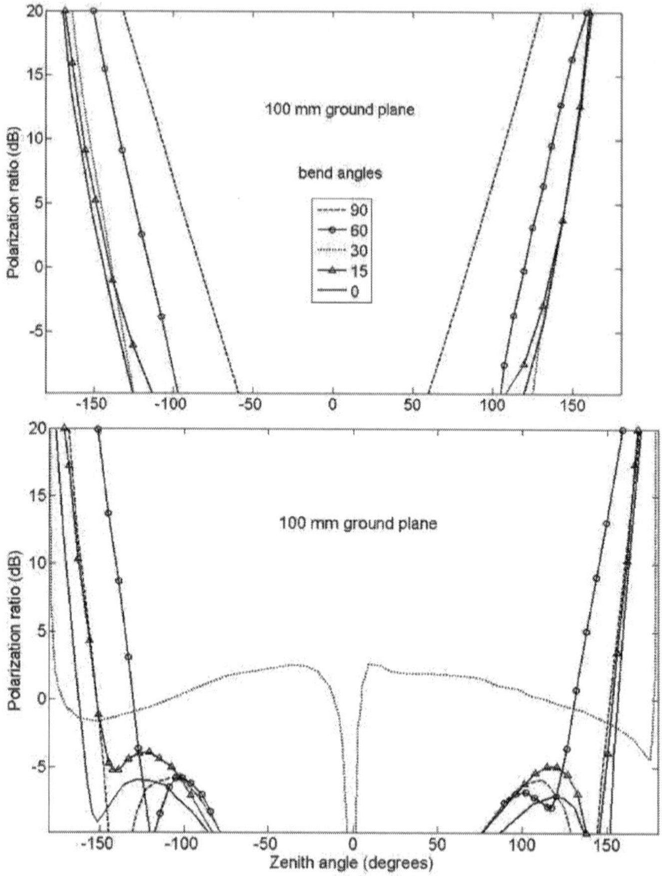

Figure 15. Cross polarization rejection (Defined as the ratio of the cross-polarized to the co-polarized components) for different bend angles (0^0, 15^0, 30^0, 60^0, and 90^0); top, $\varepsilon_r = 4.2$, bottom, $\varepsilon_r = 2.2$.

3.2. Drooped microstrip with an upward bend

The modification of the FDTD model to accommodate upward bends was accomplished by interchanging the positions of the ground plane and the element. The structure would then be upside down in the computational space but would have an upward bend. Initially, we considered the antenna shown in Figure16, which has a 30 mm flat top on a substrate with $\varepsilon_r = 2.2$. Three ground plane sizes were analyzed with bend angles varied up to 90^0. Figure 17 shows the gain and the 3-dB beamwidth for the upward bend cases. As seen, a noticeable beam broadening is evident at the higher bend angles with 3-dB beamwidths up to 60% greater than the equivalent flat case. One can observe, however, a distinct reduction in the beamwidth for the initial small bend angles, particularly for larger ground planes. The beamwidth did not recover to that of the equivalent downward case until the bend angle exceeded 60^0.

A prototype antenna was constructed and tested to allow comparison against experimental results. The antenna is identical to the one shown in Figure 13; only two variable upward bends were used in this case. For measurement purposes, a linearly polarized excitation was used. The initial dip in the beamwidth did not occur in the simulated and measured results shown in Fig. 18.

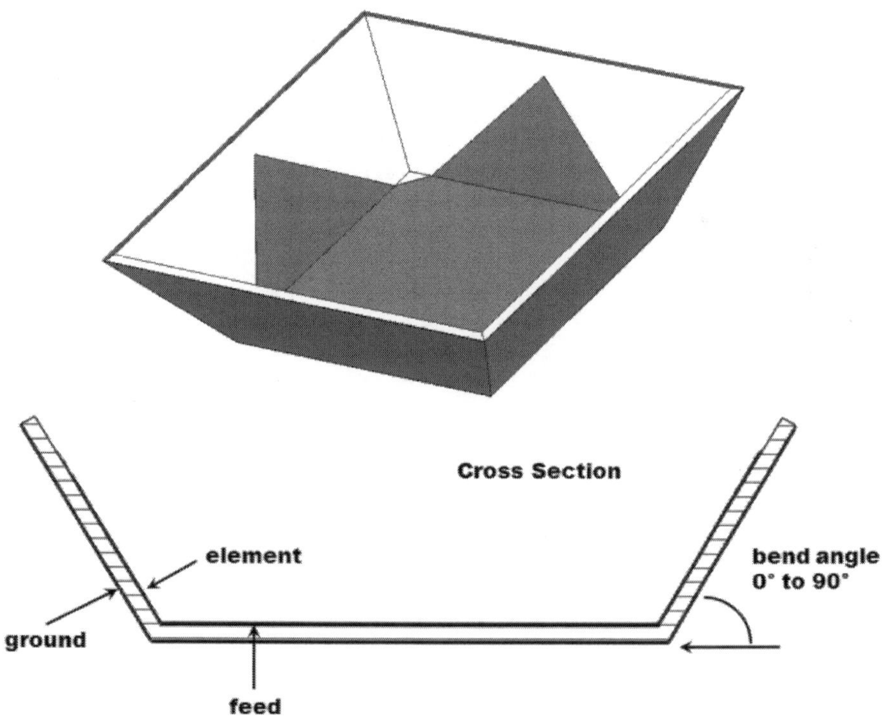

Figure 16. Geometry of the upward bend antenna.

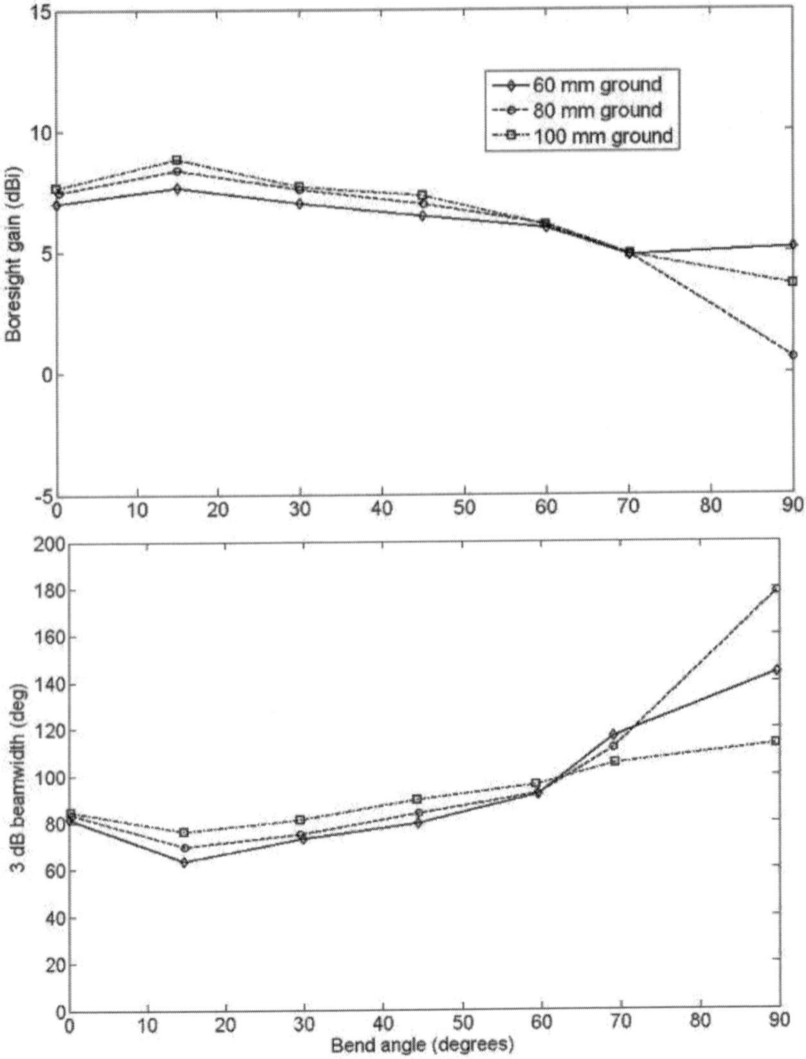

Figure 17. Gain and beam width variation with upward bend angle for three ground plane sizes, ε_r = 2.2. (Bend forming a 30 mm flat center section).

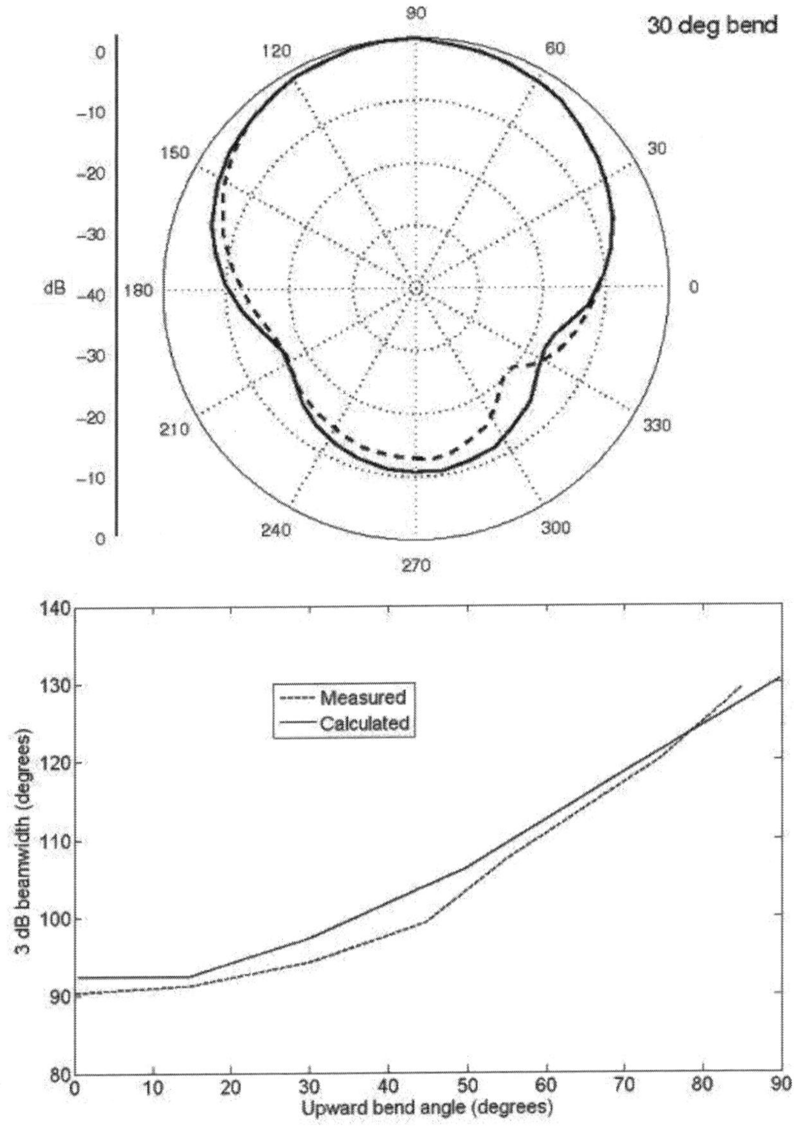

Figure 18. *E* plane elevation pattern and beamwidth behavior of the adjustable upward bend antenna.

With the completion of the initial set of simulations and measurements, we pursued the model with additional parameter variations. Bend locations at the 10 and 50 mm positions were tested for the $\varepsilon_r = 2.2$ substrate. Also, the $\varepsilon_r = 4.2$ substrate was examined when the bend was positioned at the 30 mm position. The results are summarized in Tables 4 and 5.

A changing upward bend produced little effect on the phase in the upper hemisphere. All phase curves above 90^0 elevation fell within 30^0 of each other over the full range of bend angles. Like the downward bend, the RMS phase error over the hemisphere did not vary substantially with bend angle, although a slightly greater differentiation is evident between different ground plane sizes. The elevation gain and cross polarization rejection for various upward bend angles are displayed in Figs. 19 and 20, respectively. When compared to Figure 15, it is apparent that the rejection is better by 2 dB near the horizon for upward bends, but the opposite is true below the horizon. In both cases, bending the ground plane reduces the cross-polarization discrimination near the horizon.

Table 4. Results for the drooped microstrip, substrate permittivity 2.2.

ε_r=2.2		Boresight gain (dBi)			3 dB Beamwidth (deg)			Near horizon gain Roll-off (dB)			RMS phase error (deg)		
Width-mm		60	80	100	60	80	100	60	80	100	60	80	100
Bend angle (10 mm top)	0	7.2	7.5	7.7	88	83	80	-11.	-12.	-13.	3.0	2.6	2.0
	15	4.0	6.2	6.4	158	134	118	-14.	-11.	-11.	6.2	6.8	9.1
	30	4.7	6.5	1.3	143	122	120	-9.4	-10.	-12.	3.6	3.5	5.5
	45	5.1	3.5	3.5	137	170	106	-7.2	-7.6	-9.4	4.5	3.3	3.3
	60	4.1	-.8	2.6	142	302	214	-3.8	-6.7	-7.1	6.6	5.2	4.6
	70	2.5	-4.4	-.42	127	327	300	0.44	5.8	5.7	35.8	6.5	11.2
	90	-1.5	-6.1	-6.4	188	360	360	-2.3	-6.6	0.93	11.0	69.9	80.1
Bend angle (30 mm top)	0	7.2	7.5	7.7	88	83	80	-11.	-12.	-13.	3.0	2.6	2.0
	15	7.7	8.4	8.7	75	69	63	-12.	-13.	-15.	1.8	1.6	3.9
	30	7.1	7.6	7.7	81	75	73	-11.	-12.	-13.	2.8	2.4	4.0
	45	6.6	7.0	7.3	89	83	79	-9.1	-9.8	-11.	2.3	1.5	2.0
	60	6.0	6.2	6.0	95	91	91	-8.3	-7.8	-7.9	2.2	1.3	1.5
	70	4.9	4.9	4.9	104	110	114	-6.7	-5.8	-5.3	4.4	2.8	1.9
	90	5.1	0.6	3.7	112	176	142	-6.1	-2.1	-3.9	6.8	4.7	1.5
Bend angle (50 mm top)	0	7.2	7.5	7.7	88	83	80	-11.	-12.	-13.	3.0	2.6	2.0
	15	7.3	7.6	7.9	82	75	73	-12.	-13.	-14.	1.2	1.5	2.5
	30	7.0	7.4	7.9	79	77	69	-12.	-12.	-13.	1.6	2.4	3.4
	45	7.3	7.5	7.8	83	79	73	-10.	-11.	-12.	1.1	1.5	1.9
	60	7.1	6.8	7.1	87	81	87	-9.7	-9.4	-10.	1.0	1.1	1.3
	70	6.9	6.8	6.9	83	87	81	-9.4	-5.4	-4.5	2.5	2.9	2.0
	90	6.5	5.8	5.1	89	99	110	-8.9	-7.2	-6.4	5.6	12.6	10.3

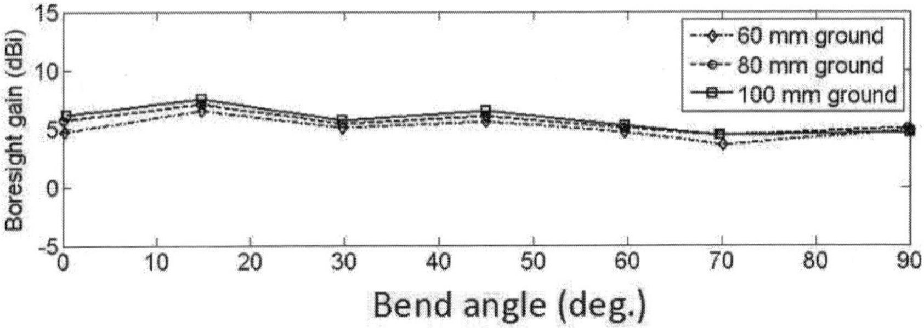

Figure 19. Gain variation with upward bend angle for three ground plane sizes, ε_r = 4.2. (Bend forming a 30 mm flat center section).

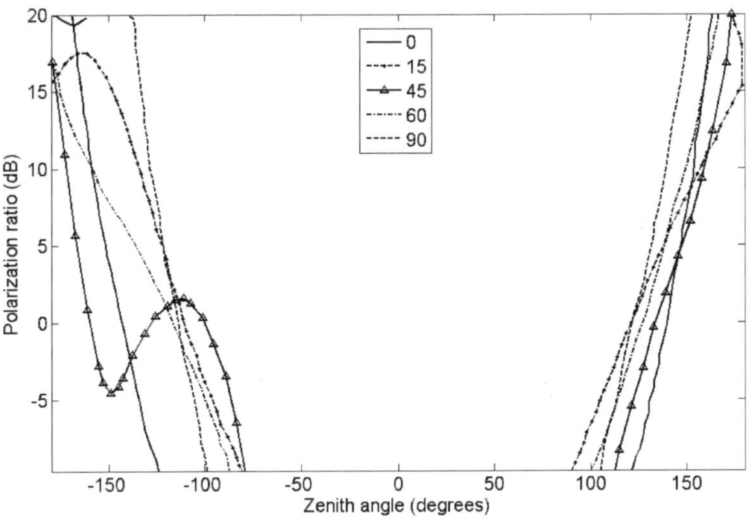

Figure 20. Cross polarization rejection (Defined as the ratio of the cross-polarized to the co-polarized components) for different bend angles (0^0, 15^0, 45^0, 60^0, and 90^0), 100 mm ground plane, ε_r = 2.2.

4. CONCLUDING REMARKS

In this chapter, we have presented numerical simulations and experimental measurements to analyze downward and upward drooped microstrip antennas with the intent of modifying the radiation pattern of the basic planar patch to accommodate the coverage requirements of GPS marine navigation and positioning. Magnitude and phase of the simulated and measured far-field radiation patterns are presented to reveal the tradeoffs in performance between patch geometry, ground plane size, and orientation. Results reported for the wide range of structural variations applied to the base antenna along with changes in the substrate material should be valuable to designers seeking to achieve a specific coverage performance.

It has been found that an accurate and stable phase center can be obtained over the entire hemisphere for moderate upward bends. Numerical simulations and measurements demonstrate that the 3-dB beamwidth of the flat microstrip patch can be increased by at least 15% and 60% for the downward and upward bends, respectively. The phase stability demonstrated by the slightly bent structures may be viewed as advantageous in cases where circumstances require distorting the element but where a significant alteration in the pattern is not desired. These were accomplished, however, at the expense of some loss of the low-profile character of the antenna.

Although not dramatically affected, the cross polarization discrimination of the bent antennas is reduced by 3 dB at the horizon compared to the equivalent flat case. Calculations of the RMS error in the spherical phase fit over the upper hemisphere showed little change with bend angle or position. In general, there is a tradeoff in achieving broad-beam pattern coverage, and maintaining high cross polarization discrimination.

The crossed dipole source, when used with the pedestal ground plane, demonstrated significant pattern improvements which inspired our interest in the drooped microstrip structure. It should be noted, however, that the crossed dipole is fundamentally different from the microstrip antenna, being itself a stand-alone radiator which operates in the presence of secondary sources created by the ground plane image. For the microstrip antenna, the ground plane is an integral part of the structure. The interior field distribution of the fundamental mode remains essentially unchanged with ground plane manipulation, and pattern modification can come about only by repositioning of the radiating edges in space.

REFERENCES

1. D Wells, Editor), Guide to GPS Positioning, Larry D Hothem, December 1986

2. C. H Ho, P. K Shumaker, L Fan, K. B Smith, and J. W Liao, Printed Cylindrical Slot Antenna for GPS Commercial Applications. *Electron. Lett.*, 32151152Feb. 1996

3. C. H Ho, P. K Shumaker, L Fan, K. B Smith, and J. W Liao, A Novel GPS Avionics Antenna. *IEEE Antennas Propagat. Lett.*, 195019531996

4. E. E Altshuler, Hemispherical Coverage Using a Double-Folded Monopole. *IEEE Trans. Antennas Propagat.*, 44811121119Aug. 1996

5. Z Ying, and P. S Kildal, Improvement of Dipole, Helix, Spiral, Microstrip Patch and Aperture Antennas with Ground Planes by Using Corrugated Soft Surfaces. *IEE Proc. Microw. Antennas Propagat.*, 1433244248Jun. 1996

6. D. M Pozar, and S. M Duffy, A Dual-band, Circularly Polarized Aperture-Coupled Stacked Microstrip Antenna for Global Positioning Satellite. *IEEE Trans. Antennas Propagat.*, 451116181625Nov. 1997

7. N Padros, J. I Ortigosa, J Baker, M. F Iskander, and B Thornberg, Comparative Study of High-Performance GPS Receiving Antenna Designs. *IEEE Trans. Antennas Propagat.*, 454698706Apr. 1997

8. H. K Kan, and R. B Waterhouse, A Small, CP-Printed Antenna Using 120 deg Sequential Rotation. *IEEE Trans. Antennas Propagat.*, 503398399Mar. 2002

9. S Lee, J Woo, M Ryu, and H Shin, Corrugated Circular Microstrip Patch Antennas for Miniaturization. *Electron. Lett.*, 38262263Mar. 2002

10. J. L Boccia, G Amendola, and G. Di Massa. A Shorted Elliptical Patch Antenna for GPS Applications. *IEEE Antennas Wireless Propagat. Lett.*, 2682003

11. Ozgun, S Mutlu, M. I Aksun, and L Atalan, Design of Dual-Frequency Probe-Fed Microstrip Antennas with Genetic Optimization Algorithm. *IEEE Trans. Antennas Propagat.*, 51819471954Aug. 2003

12. L Boccia, G Amendola, and G. D Massa, A Dual Frequency Microstrip Patch Antenna for High-Precision GPS Applications. *IEEE Antennas Propagat. Lett.*, 31571602004

13. K Oh, B Kim, and J Choi, Novel Integrated GPS/RKES/PCS Antenna for Vehicular Application. *IEEE Microwave Wireless Comp. Lett.*, 154244246Apr. 2005

14. L. I Basilio, J. T Williams, D. R Jackson, and M. A Khayat, A Comparative Study of a New GPS Reduced-Surface-Wave Antenna. *IEEE Antennas Propagat. Letters*, 42332362005

15. Y Zhang, and H. T Hui, A Printed Hemispherical Helical Antenna for GPS Receivers. *IEEE Microwave and Wireless Comp. Lett.*, 1511012January 2005

16. R Azaro, F De Natale, M Donelli, E Zeni, and A Massa, Synthesis of a Prefractal Dual-Band Monopolar Antenna for GPS Applications," *IEEE Antennas Wireless Propagat. Lett.*, 53613642006

17. Y Zhou, S Koulouridis, G Kiziltas, and J. L Volakis, A Novel, Quadruple Antenna for Tri-Band GPS Applications. *IEEE Antennas Wireless Propagat. Lett.*, 52242272006

18. L. I Basilio, R. L Chen, J. T Williams, and D. R Jackson, A New Planar Dual-Band GPS Antenna Designed for Reduced Susceptibility to Low-Angle Multipath. *IEEE Trans. Antennas Propagat.*, 55823582366Aug. 2007

19. L Boccia, G Amendola, and G. D Massa, Performance Evaluation of Shorted Annular Patch Antennas for High-Precision GPS Systems. *IET Microwave Antennas Propagat.*, 124654712007

20. Y Zhou, C. C Chen, and J. L Volakis, Dual Band Proximity-Fed Stacked Patch Antenna for Tri-Band GPS Applications. *IEEE Trans. Antennas Propagat.*, 551220223Jan. 2007

21. G Lachapelle, M Casey, R. M Eaton, A Kleusberg, J Tranquilla, and D Wells, GPS Marine Kinematic Positioning Accuracy and Reliability. *The Canadian Surveyor*, 412143172Summer 1987

22. G Lachapelle, C Liu, G Lu, M. E Cannon, B Townsend, and R Hare, Precise Marine DGPS Positioning Using P Code and High Performance C/A Code Technologies. *National Technical Metting, ION*, San Francisco, Jan. 1993

23. I Yaesh, and B Priel, Design of Leveling Loop for Marine Navigation System. *IEEE Trans. Aerospace Electr. Sys.*, 292599604Apr. 1993

24. D Deifes, GPS Based Attitude Determining System for Marine Navigation. *IEEE Position, Location, and Navigation Symposium*, 806812Apr. 1994

25. G Lachapelle, M. E Cannon, G Lu, and B Loncarevic, Ship borne GPS Attitude Determination During MMST-93. *IEEE J. Oceanic Eng.*, 211100105Jan. 1996

26. P. K Shumaker, C. H Ho, and K. B Smith, Printed Half-Wavelength Quadrifilar Helix Antenna for GPS Marine Applications. *Electron. Lett.*, 32153154February 1996

27. S Beiter, R Poquette, B. S Filipo, and W Goetz, Precision Hybrid Navigation System for Varied Marine Applications. *IEEE Position, Location, and Navigation Symposium*, 316323April 1998

28. J. M Tranquilla, J. P Carr, and H. M Al-rizzo, Analysis of a Choke Ring Ground Plane for Multipath Control in Global Positioning System (GPS) Applications. *IEEE Trans. Antennas Propagat.*, 42905911Oct. 1994

29. C. C Counselman, III. Multipath-Rejecting GPS Antennas. *Proc. of the IEEE*, 8718691Jan. 1999

30. B. R Townsend, and P. C Fenton, A Practical Approach for the Reduction of Pseudorange Multipath Errors in a L1 GPS Receiver. ION-94, Salt Lake City, 16Sept. 1994

31. J. M Tranquilla, The Experimental Study of Global Positioning Satellite Antenna Backplane Configurations. Technical Report, NASA Jet Propulsion Lab., Contract 957959, Radiating Systems Research Laboratory, University of New Brunswick, Fredericton, NB, Canada, 1988

32. J. M Tranquilla, B. G Colpitts, and J. P Carr, Measurement of Low Multipath Antennas for Topex. Proceedings of the *5th International Geodetic Symposium on Satellite Positioning*, Las Cruces, NM, 356361Mar. 13-1 7, 1989

33. J. M Tanquilla, and B. G Colpitts, Development of a Class of Antennas for Space-Based NAVSTAR GPS Applications. *International Conference on Antennas and Propagation (ICAP 89)*, 6th, Coventry, England, Apr. 4-7, 1989, Proceedings. Part 1 (A90-27776 11-32). London, England and Piscataway, NJ, Institution of Electrical Engineers, 65691989

34. J. M Tranquilla, and B. G Colpitts, GPS Antenna Design Characteristics for High Precision Applications. Proceedings of the *ASCE Conference GPS-88 Engineering Applications of GPS Satellite Surveying Technology*, Nashville, TN, May 11141988

35. J. M Tranquilla, and S. R Best, A Study of the Quadrifilar Helix Antenna for Global Positioning Systems (GPS) Applications. *IEEE Trans. Antennas Propagat.*, 381545155OOct.1990

36. S Noghanian, and L Shafai, Control of Microstrip Antenna Radiation Characteristics by Ground Plane Size and Shape. *IEE Proc. Microw. Antennas Propagat.*, 145207212Ju.1998

37. K. G Clark, The Finite-Difference Time-Domain Technique Applied to the Drooped Microstrip Antenna. PhD thesis, Department of Electrical Engineering, University of New Brunswick, Fredericton, NB, Canada, Jul. 1996

38. W Feller, Three Dimensional Microstrip Patch Antenna. US Patent Publication 5Apr. 1993

39. C. W Su, S. K Huang, and C. H Lee, CP Microstrip Antenna with Wide Beamwidth for GPS Band Application. *Electronics Lett.*, 4320th Sept. 2007

40. B. G Colpitts, The Uniform Theory of Diffraction Applied to Wedges and Curved Surfaces. PhD thesis, Department of Electrical Engineering, University of New Brunswick, Fredericton, NB, Canada, Apr.1988

41. A Taflove, and K. R Umashankar, Finite-Difference Time-Domain Modeling of Electromagnetic Wave Scattering and Interaction Problems. *IEEE Trans. Antennas Propagat. Society Newslett.*, 520Apr. 1988

42. T. G Jurgens, A Taflove, K. R Umashankar, and T. G Moore, Finite-Difference Time-Domain Modeling of Curved Surfaces. *IEEE Trans. Antennas Propagat.*, AP-40357366Apr. 1992

43. J. M Tranquilla, and S. R Best, Phase Center Considerations for the Monopole Antenna. *IEEE Trans. Antennas Propagat.*, AP-34741744May 1986

9

CHAPTER

RECONFIGURABLE MICROSTRIP DOUBLE-DIPOLE ANTENNAS FOR PERSONAL WIRELESS COMMUNICATIONS

Abdelnasser Eldek, Abubaker Abdallah, Mahmoud Manzoul

Department of Computer Engineering, Jackson State University, Jackson, USA.

ABSTRACT

In this paper, a number of reconfigurable antennas are designed to support multiband and wideband wireless applications in different frequency bands in the range between 1 GHz and 6 GHz. The proposed designs consist of microstrip-fed two dipoles through simplified balun. By controlling the lengths of the dipoles by connecting or disconnecting slots, we are able to change the antenna operating frequency as well as the operation mode. Two modes are supported by all of the proposed antennas: Multiband and Broadband modes.

Keywords: Reconfigurable Antennas, Dipole, Wideband, Multiband

1. INTRODUCTION

As communication devices become smaller due to greater integration of electronics, the antenna becomes a significantly larger part of the overall package volume. This results in a demand for similar reductions in antenna size. However, reducing antenna size without significantly impacting gain and efficiency is a challenging task. As integration increases, a single antenna is often required to support two or more of the many wireless services across a broad frequency range. Multiband, wideband and reconfigurable antennas are being developed to meet this need [1]. Multi-band and wideband antennas require complicated filters with stringent requirements to improve their out-of-band noise-rejection performance. These filters, moreover, are generally bulky and expensive. In contrast, frequency-reconfigurable antennas have great potential for reducing production cost and offer better out-of-band noise rejection [2]. In addition, compared to multiband and broadband antennas, reconfigurable antennas offer the advantages of compact size, similar radiation pattern for all designed frequency bands, efficient use of electromagnetic spectrum and frequency selectivity useful for reducing the adverse effects of co-site interference and jamming [3]. Dual frequency reconfigurable microstrip antennas can offer additional advantages of frequency reuse for doubling the system capability and polarization diversity for good performance of reception and transmission or to integrate the receiving and transmitting functions into one antenna for reducing the antenna size [4,5].

Microstrip antennas are attractive candidates for reconfigurable antenna technology because of their inherent low profile, light weight, low cost, and ease of fabrication and installation [2,6-13]. Number of attractive reconfigurable antenna designs were presented in [2] with the objective of reduction of the number of MEMs utilizing antenna symmetry to decrease the cost. The concept of a frequency-reconfigurable rectangular ring slot antennas fed by slotlines or CPW is presented in [7]. A reconfigurable patch antenna is obtained in [8] by inserting slits at the non-radiating edges of the patch. A reconfigurable Yagi antenna is presented in [9] to operate at 2.4 and 5.78 GHz for wireless communications. A reconfigurable slot dipole antenna was presented in [10] in the X-band. Patch antenna with polarization diversity using switchable L-shaped slots is presented in [11]. A reconfigurable dipole antenna with harmonic trap was presented in [12] where the operating frequency is controlled by a reconfigurable dipole length, while the higher-order modes are eliminated using a reconfigurable l/4 open circuit stub. A compact multiband antenna has been redesigned to be reconfigurable with embedded switches to allow selective addressing of these bands [13].

In this paper, three microstrip-fed double-dipole antenna designs [14-16] to support multi-band, wideband, and single-band operation modes are presented for personal wireless communications. The reconfiguration for the microstrip antennas is carried out by connecting or disconnecting (switching on or off) appropriate rectangular strips to model the MEM switches. The shape and number of these strips along with the switching state and position is used to control the operating frequency of the antenna without changing the feeding mechanism. The antennas' geometries and their results will be explained in the next two sections. The full-wave electromagnetic simulations and analysis for the presented antenna are performed using the commercial computer software package Ansoft High Frequency Structure Simulator (HFSS) [17], which is based on the finite element method. Measurements of the return loss and radiation patterns are also conducted to verify the simulation results and demonstrate the feasibility of the proposed configurations.

2. ANTENNAS' GEOMETRIES AND PARAMETRIC STUDY

Three double-dipole antennas were designed for different frequency bands and depicted in Figure 1. The double-dipole antenna consists of two dipoles with similar or different lengths depending on the required operating frequencies. A simplified balun is used to feed the dipoles. In each dipole, the left half is printed on the top substrate layer and fed by a microstrip line, while the right half is printed on the bottom layer and connected to the truncated ground plane.

The length of the dipole is approximately half the wavelength at the operating frequency, while its distance from the truncated ground plane is quarter wavelength for the best performance. Thus, if the antenna is required to operate at a lower frequency (f_L) and a higher frequency (f_H), then the shorter dipole will be closer to the ground plane. In this case, the antenna will operate at multiple bands. A broadband design can be achieved by making the longer and the shorter dipoles to resonate at closer frequencies. This can be achieved by reversing the positions of the two dipoles. With the contribution of the coplanar strip (CPS) line length to the antenna length, the difference between their lengths become small, and the two operating bands will be close and can merge in one relatively wide frequency band.

The next step is to merge these two modes in one so that the length of the closer dipole is controlled by the lower operating frequency in the wideband mode, and the length of the second dipole is controlled by the higher operating frequency in the multiband mode. Next, a proper matching circuit between the two dipoles is designed to eliminate the mutual effect between the two modes. Therefore, the shape of the CPS connecting the two dipoles is modified as needed by the design to obtain the required modes. Finally, when adding the ground plane, it will have an effect on the performance; therefore, the last step is to shape the ground plane to reduce this effect as shown in Figure 1. A smooth transition from the microstrip line mode to the two CPS line mode is used for this purpose.

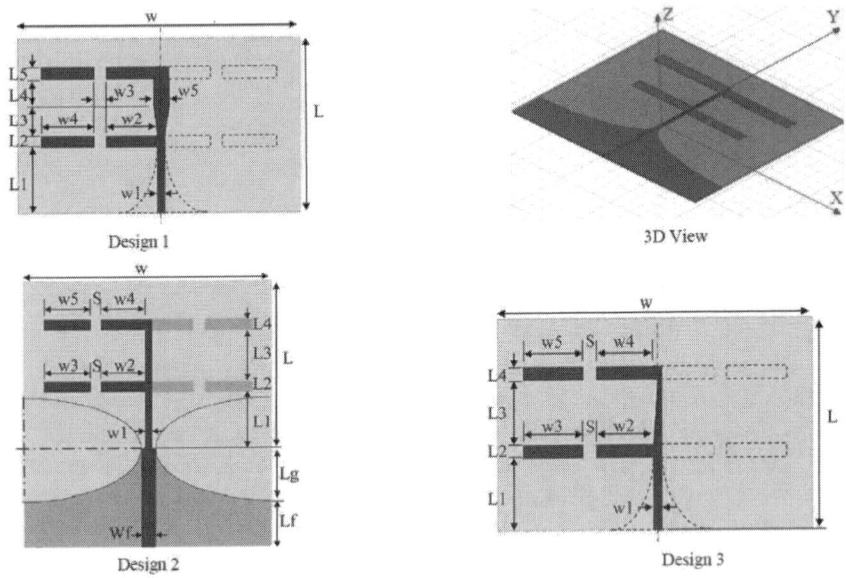

Figure 1. Antennas' geometries and paramters.

Parametric study is performed on each antenna and depicted in Figures 2 to 11 for Design 1. This study is done in the multiband mode to know the effect of each parameter, and select the best dimensions, and tune the antenna to obtain the best bandwidth in the broadband mode without affecting the multiband mode. The dimensions of the three designs resulted from the parametric study are shown in Table 1.

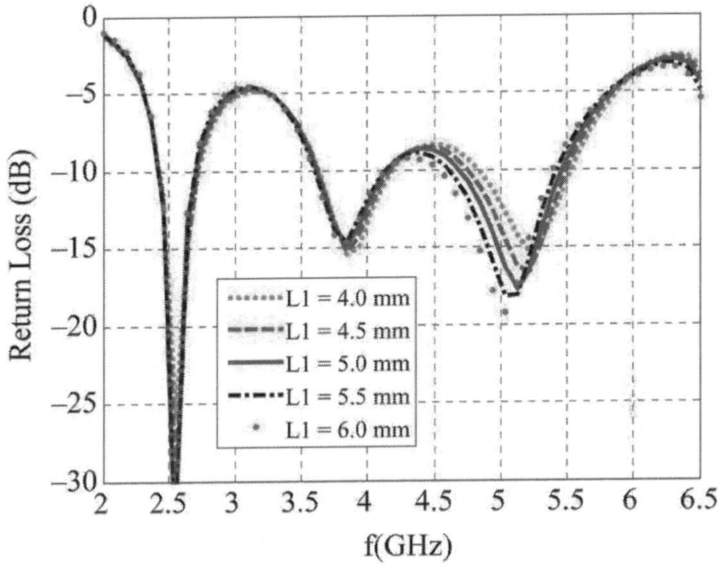

Figure 2. Effect of L1.

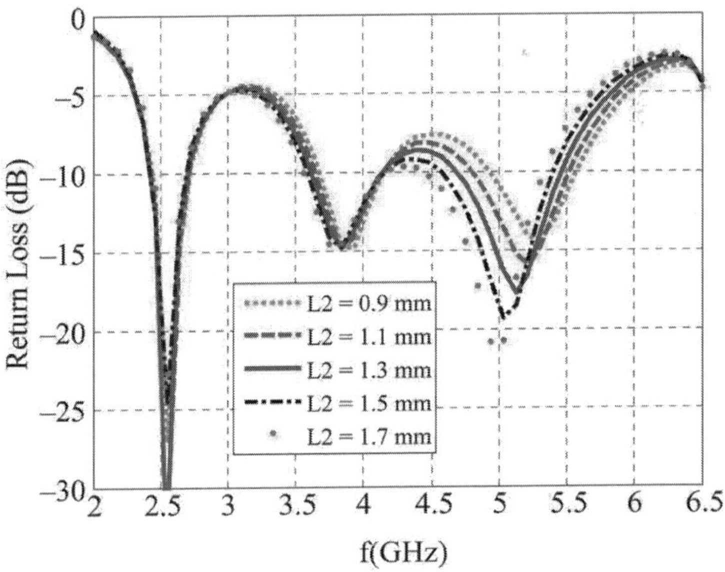

Figure 3. Effect of L2.

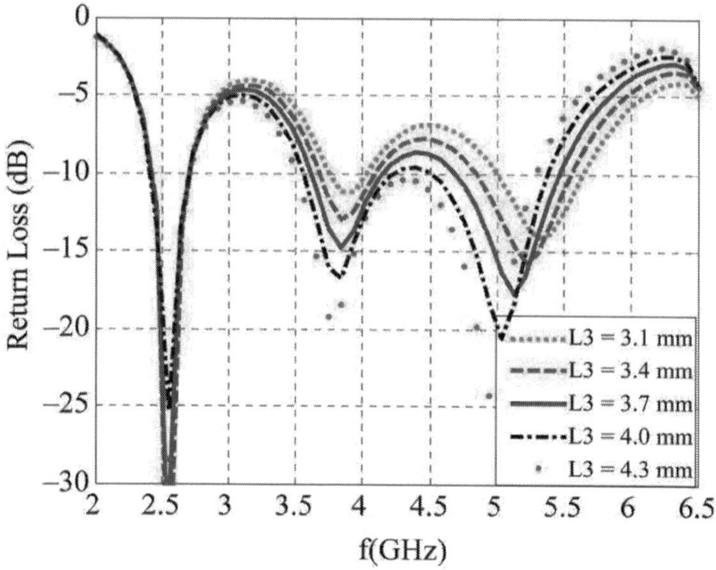

Figure 4. Effect of L3.

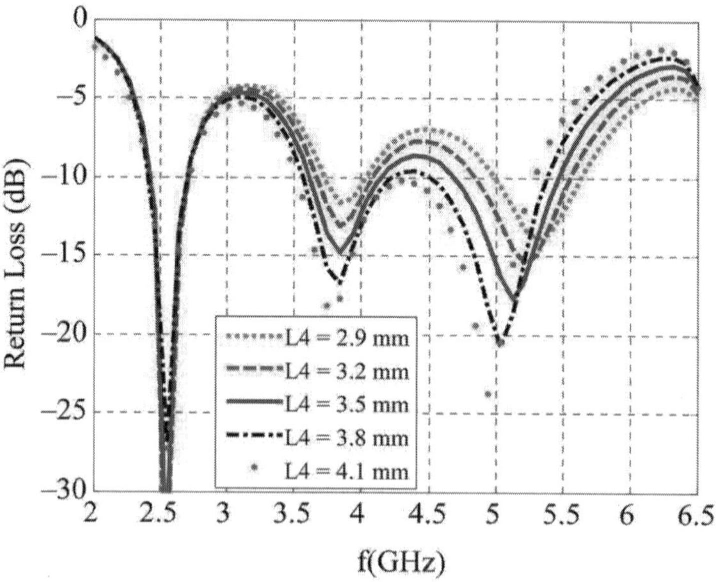

Figure 5. Effect of L4.

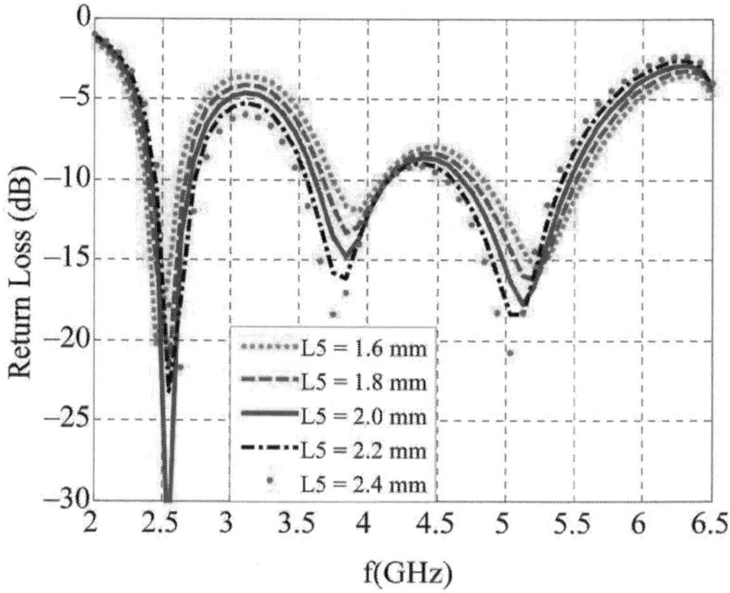

Figure 6. Effect of L5.

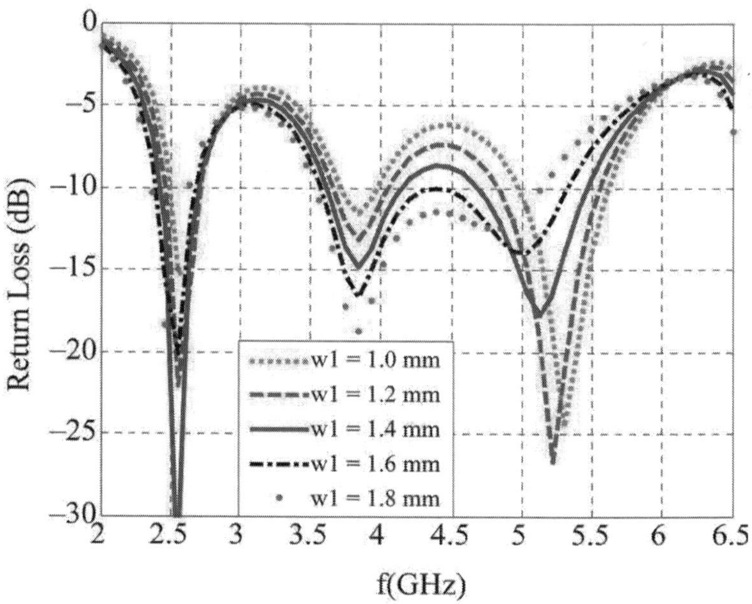

Figure 7. Effect of w1.

3. RESULTS

The return loss and radiation patterns are computed using Ansoft HFSS for Designs 1, 2 and 3. The reconfiguration is carried out by connecting or disconnecting appropriate rectangular strips to model the MEM switches. The rectangle width is "S" and its height is the same like the strip on which Pin diode or MEMs can be installed. Since we have four switches, we can create 16 cases starting from Case 0000 when all switches are open, to Case 1111 when all switches are closed. However, only 10 different combinations are considered after excluding the similar cases such as 0010 and 0001.

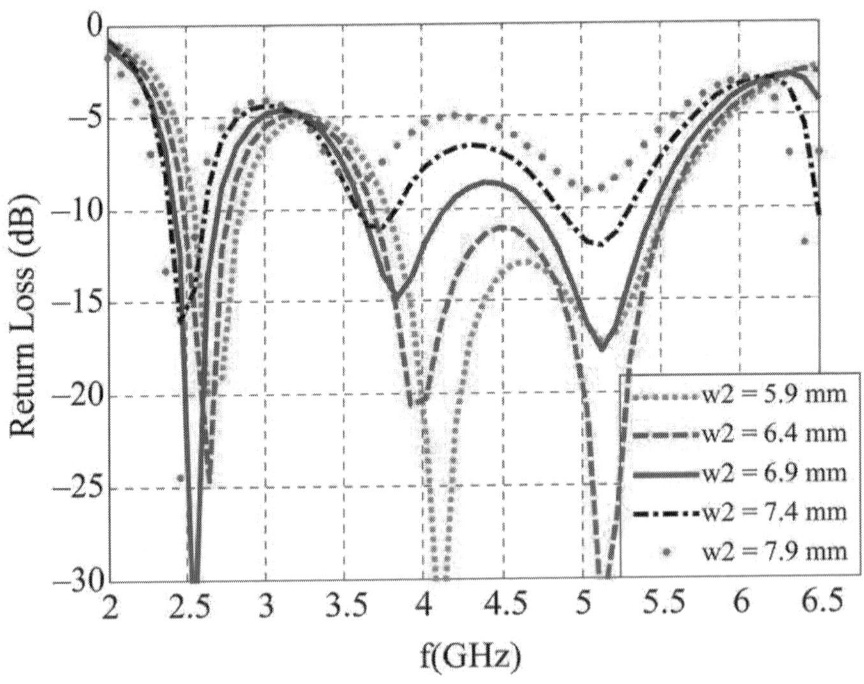

Figure 8. Effect of w2.

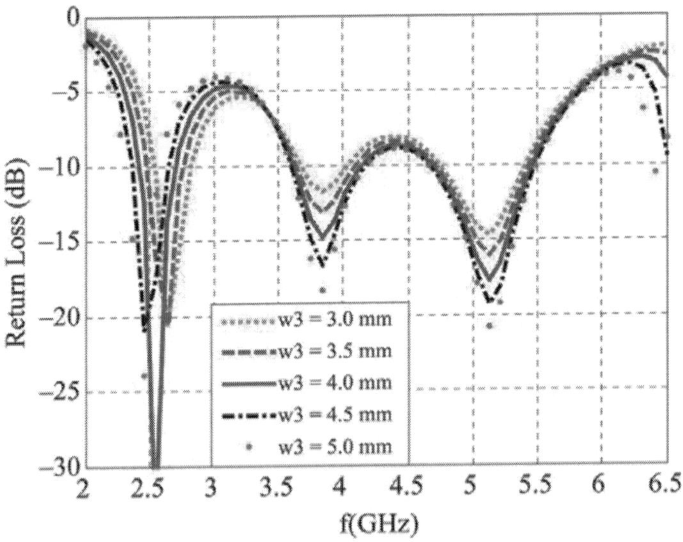

Figure 9. Effect of w3.

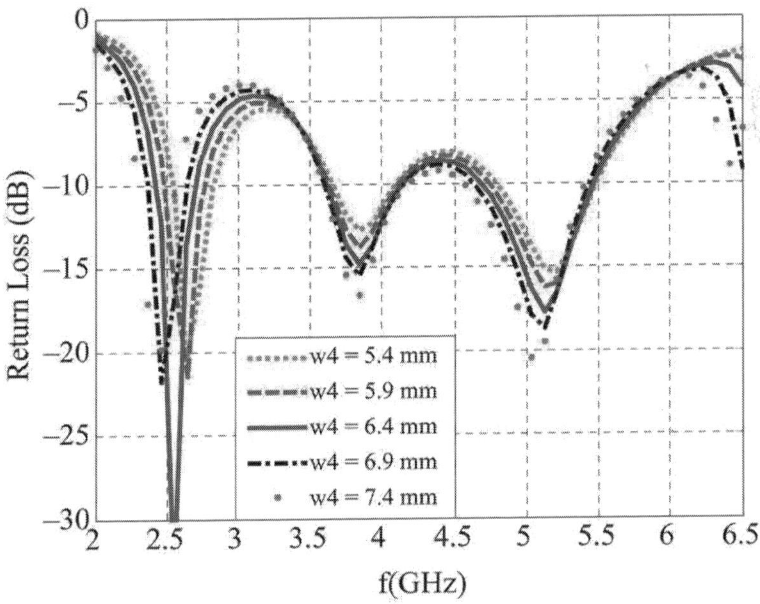

Figure 10. Effect of w4.

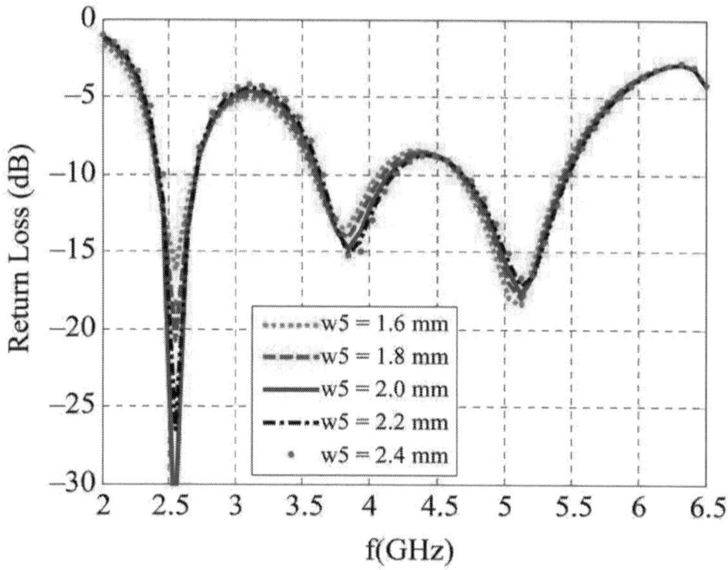

Figure 11. Effect of w5.

Table 1. Dimensions (mm) of the proposed designs.

	Design 1	Design 2	Design 3
W	46.2	93.5	144.5
L	18.9	46.3	89
w1	0.6	1.3	3.5
w2	7	9	22
w3	1	16.8	31
w4	9.8	19.8	32
w5	1.2	12.8	30
L1	5	12	42
L2	1.3	4.3	3
L3	3.1	13	36
L4	2.9	5	3
L5	1.6	N/A	N/A
S	N/A	1	1

Figure 12 shows the multiband (0011) and broadband (1100) cases in addition to the cases when all switches are ON/OFF (0000 and 1111). These cases are the symmetrical cases where S1 and S3 are the same as S2 and S4, respectively. The operating bands of the three designs in this figure are summarized in Tables 2, 3, and 4 for Designs 1, 2, and 3, respectively. With only four cases out of ten, Design 1 covers 90% of the band between 2.22 and 6.28 (95% bandwidth), Design 2 can operate in 90% of the band between 1.64 and 3.2 (64% bandwidth), and Design 3 also covers almost all the band between 0.83 and 2.34 (95% bandwidth). In the broadband Case, the three Designs provide from 22% to 56% bandwidth. Figure 13 shows some unsymmetrical cases which show that we still can make these antennas operate at other frequencies.

These results show many advantages for these designs. First, we can switch between different frequency bands without changing the antenna. Second, we can also change between multi-band and broadband modes. In addition, the overall antenna operating bandwidth is at least 1.7 times the maximum bandwidth of all cases.

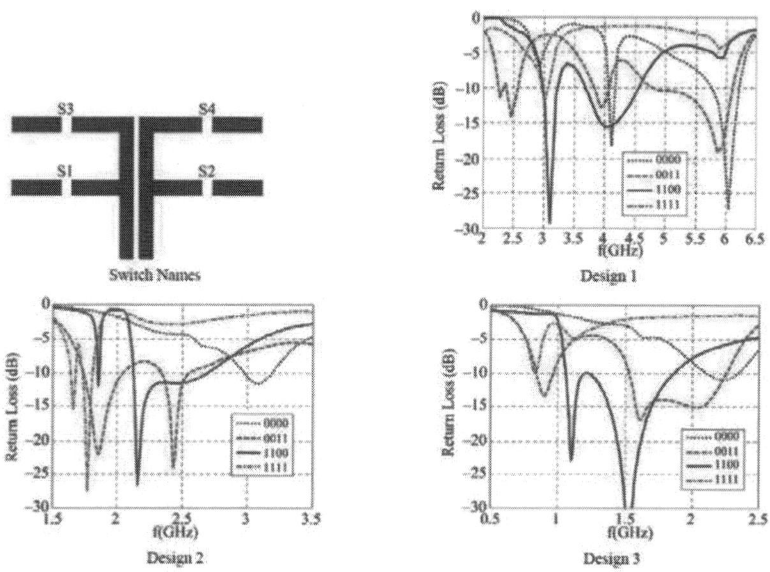

Figure 12. Return Loss (dB) for the symitrical cases.

Table 2. Operating frequency bands and bandwidth for Design 1 in the symitrical cases.

	Case	Band 1	Band 2	Band 3
	0000	4.02 - 4.15 (3%)	5.77 - 6.28 (8.5%)	N/A
Design No. 1	0011	2.22 - 2.55 (14%)	3.82 - 4.05 (6%)	4.75 - 6.10 (25%)
	1100	2.96 - 3.25 (9%)	3.65 - 4.55 (22%)	N/A
	1111	2.95 - 3.10 (5%)	N/A	N/A

Table 3. Operating frequency bands and bandwidth for Design 2 in the symitrical cases.

	Case	Band 1	Band 2
	0000	2.95 - 3.20 (8%)	N/A
Design No.2	0011	1.73 - 2.05 (17%)	2.36 - 2.53 (7%)
	1100	1.85 - 1.86 (0.5%)	2.13 - 2.73 (24%)
	1111	1.64 - 1.68 (2.4%)	1.73 - 1.80 (4%)

Table 4. Operating frequency bands and bandwidth for Design 3 in the symitrical cases.

	Case	Band 1	Band 2
	0000	2.1 - 2.34 (11%)	N/A
Design No.3	0011	0.83 - 0.98 (17%)	1.52 - 2.25 (39%)
	1100	1.05 - 1.86 (56%)	N/A
	1111	0.83 - 0.84 (1%)	N/A

Prototypes from the three designs are fabricated using LPKF milling machine. The return loss is measured using vector network analyzer for the different reconfigurable cases. Figure 14 shows a good agreement between the measured and computed results in the multi-band (MB) and broadband (BB) modes. Finally, the radiation patterns are computed and measured using anechoic chamber. Figures 15 to 17 show the co-polarized fields at selected resonance frequencies for Designs 1, 2 and 3, respectively, for the MB and BB cases. The lower half is cropped because of the symmetry. An acceptable match is obtained between the measured and computed results. Currently, we are in the process of sending antenna prototypes to specialized for integrating reconfigurable enabling technology such as MEMs in these designs.

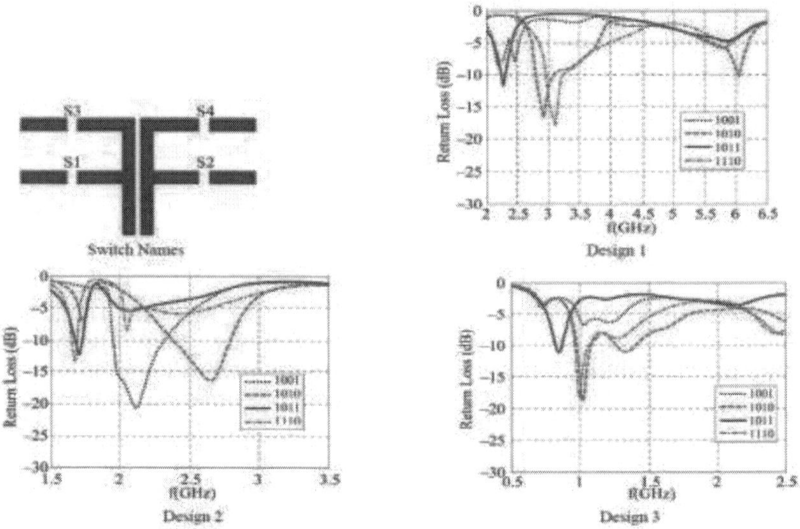

Figure 13. Return Loss (dB) for the unsymitrical cases.

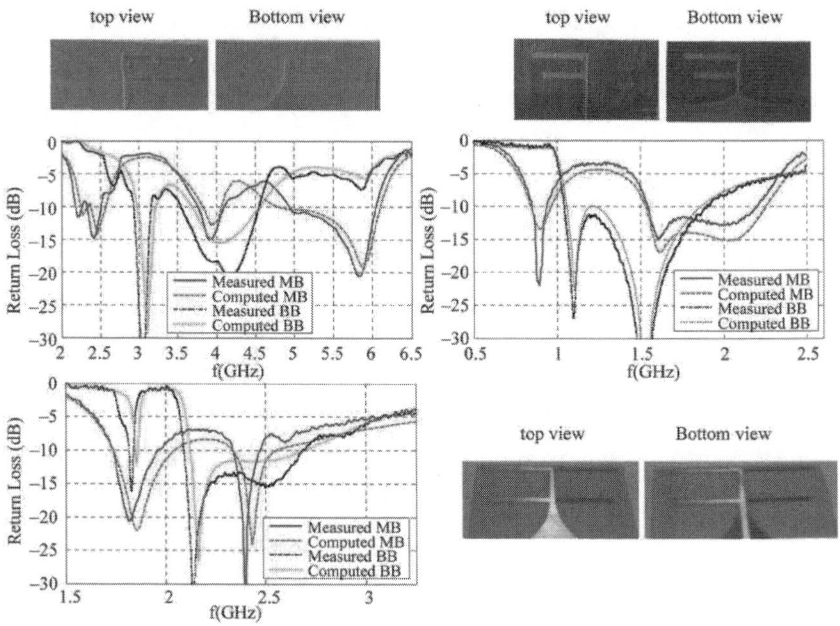

Figure 14. Measured and computed return loss and antenna prototypes.

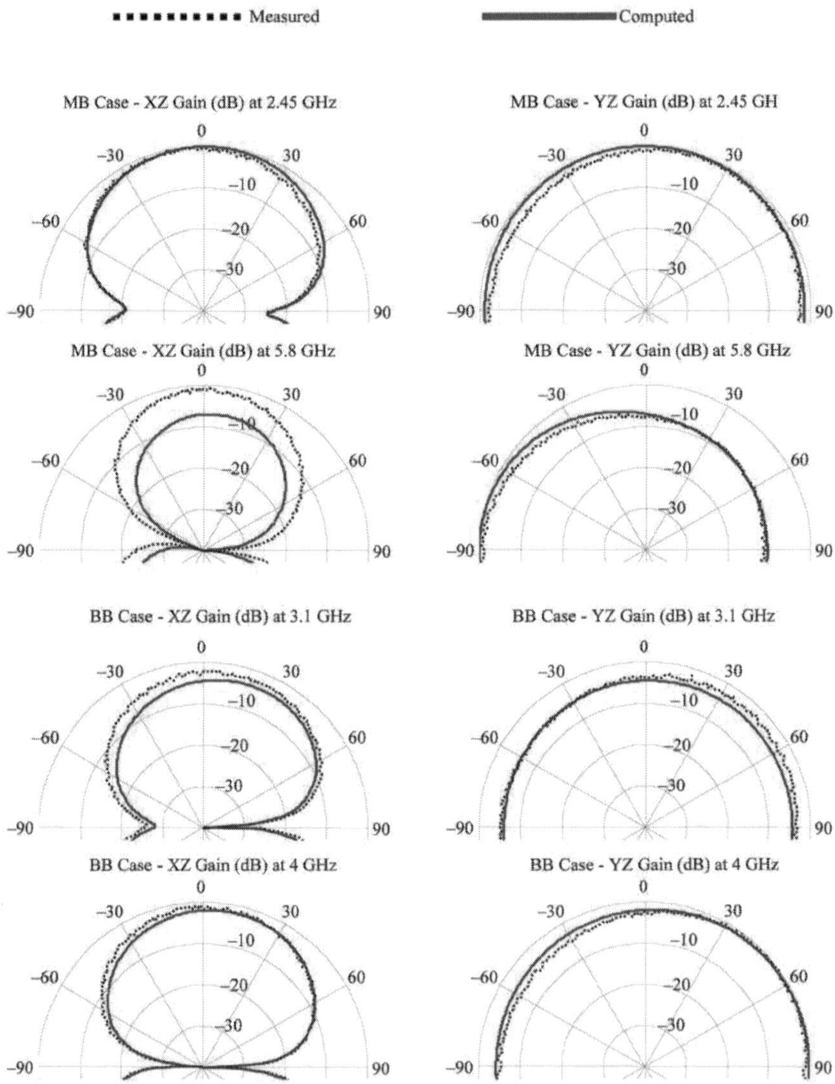

Figure 15. Radiation Patterns for Design 1 in multiband and broadband modes.

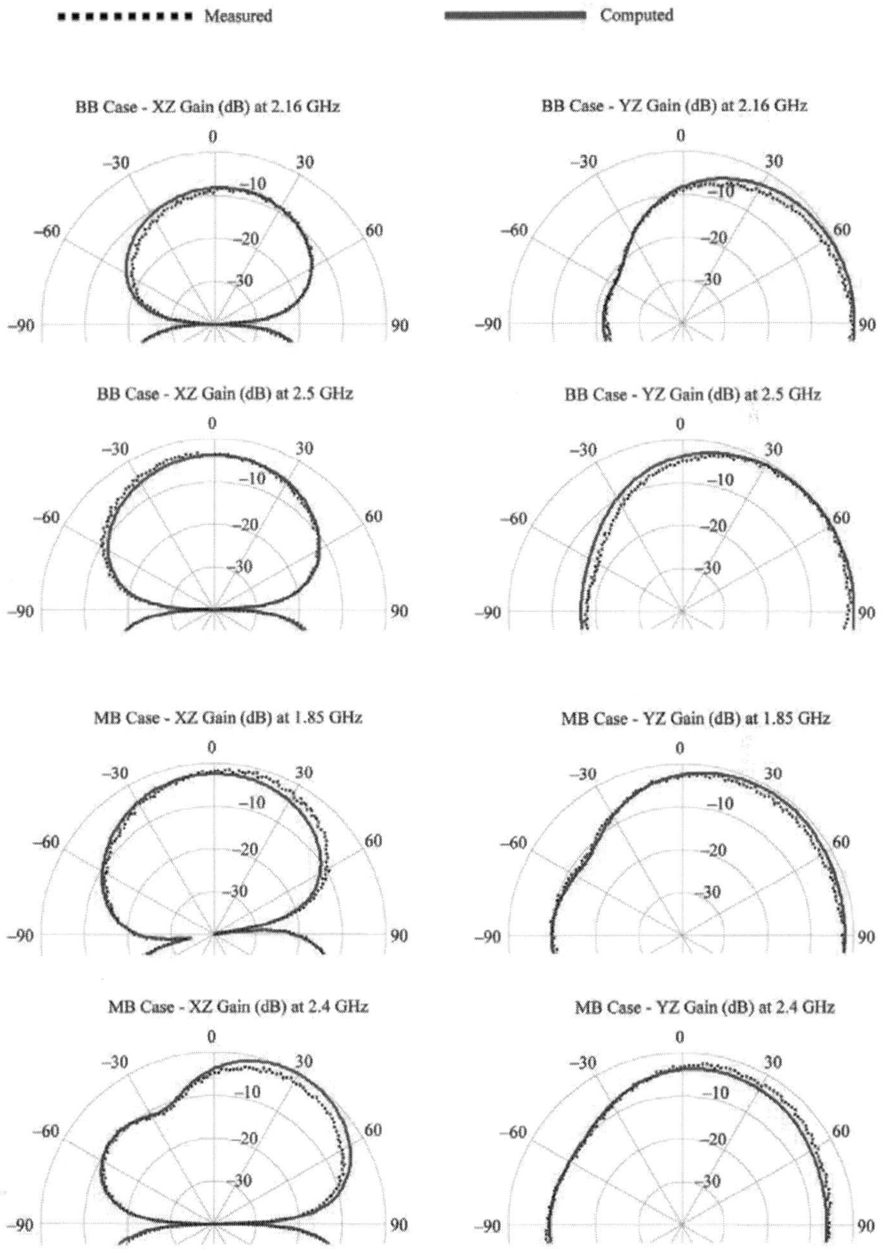

Figure 16. Radiation Patterns for Design 2 in multiband and broadband modes.

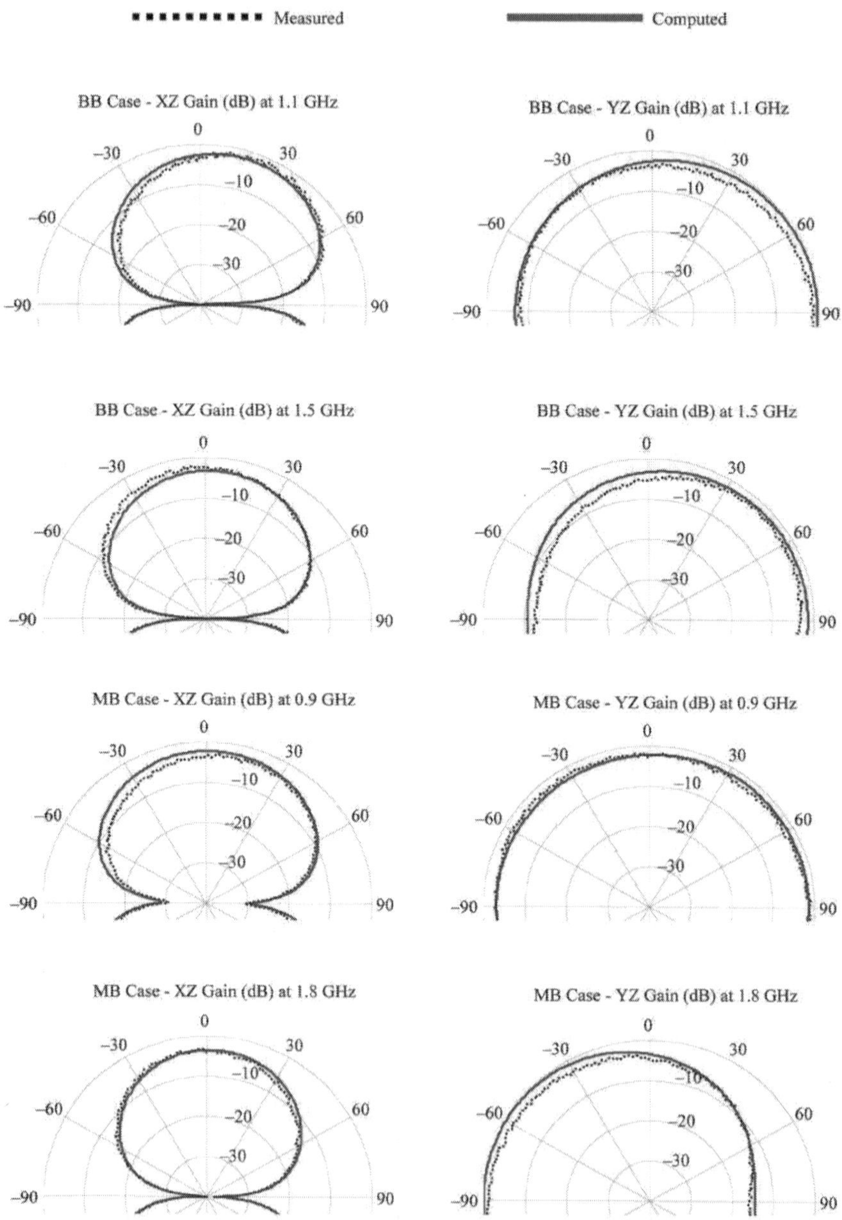

Figure 17. Radiation Patterns for Design 3 in multiband and broadband modes.

4. CONCLUSIONS

Three novel reconfigurable microstrip antenna designs are presented for wireless communications applications. The proposed antennas are combination of two dipole antennas of similar and different lengths. By changing the dipoles' lengths, we were able to change the antenna operating frequency as well as the operation mode. The antenna is merging two operating modes: Multi-band and Broadband in a single design. The presented designs cover different frequency bands: 0.83 - 2.34 GHz, 1.64 - 3.2 GHz, and 2.22 - 6.28 GHz. A wide overall bandwidth of 95% was obtained and a wide single case operating bandwidth of 56% was also provided.

5. ACKNOWLEDGEMENTS

The authors would like thank the Air Force Research Laboratory (AFRL)—Minority Leaders Program (MLP), and Clarkson Aerospace Corp for funding this research through contract number FA8650-05-D-1912. Many thanks are also to Mr. William Munn and Mr. Christopher Munn from Jackson State University for their help in building the proposed antennas.

REFERENCES

1. C. B. Dietrich, Jr., R. M. Barts, W. L. Stutzman and W. A. Davis, "Trends in Antennas for Wireless Communications," Microwave Journal, vol. 46, no. 1, January 2003, p. 22.

2. S. N. Yang, C. N. Zhang, H. K. Pan, A. E. Fathy and V. K. Nair, "Frequency Reconfigurable Antennas for Multiradio Wireless Platforms," IEEE Microwave Magazine, vol. 10, no. 1, February 2009, pp. 66-83.

3. D. Peroulis, K. Sarabandi and P. B. K. Katehi, "Design of Reconfigurable Slot Antennas," IEEE Transactions on Antennas and Propagation, Vol. 53, No. 2, 2005, pp. 645- 654.

4. S. V. Shynu, G. Augustin, C. K. Aanandan, P. Mohanan, and K. Vasudevan, "A Reconfigurable Dual Frequency Slot Loaded Microstrip Antenna Controlled by Pin-Diodes," Microwave and Optical Technology Letters, vol. 44, No. 4, 2005, pp. 374-376.

5. S. V. Shynu, G. Augustin, C. K. Aanandan, P. Mohanan, and K. Vasudevan, "Design of Compact Reconfigurable Dual, Frequency Microstrip Antennas Using Varactor Diodes," Progress in Electromagnetics Research, Vol. 60, 2006, pp. 197-205.

6. W. H. Weedon, W. J. Payne and G. M. Rebeiz, "MEMsSwitched Reconfigurable Antennas," IEEE Antennas and Propagation Society International Symposium, Boston, vol. 3, 8-13 July 2001, pp. 654-657.

7. K. C. Gupta, J. Li, R. Ramadoss and C. J. Wang, "Design of Frequency-Reconfigurable Rectangular Slot Ring Antenna," IEEE Antennas and Propagation Society International Symposium, Salt Lake City, vol. 1, 16-21 July 2000, p. 326.

8. S. Xiao, B. Z. Wang and X. S. Yang, "A Novel Frequency-Reconfigurable Patch Antenna," Microwave and Optical Technology Letters, vol. 36, no. 4, February 2003, pp. 295-297. doi:10.1002/mop.10746

9. P. F. Wahid, M. A. Ali and B. C. Deloach, "A Reconfigurable Yagi Antenna for Wireless Communications," Microwave and Optical Technology Letters, vol. 140, no. 2, July 2003, pp. 140-141.

10. A. A. Eldek, A. Z. Elsherbeni, C. E. Smith and K. F. Lee, "Wideband Slot Antennas for Radar Applications," Proceedings of IEEE Radar Conference, Huntsville, 5-8 May 2003, pp. 79-84.

11. S. Raghavan, T. Shanmuganantham and M. S. K. Kumar, "Reconfigurable Patch Antenna with Switchable L-Shaped Slots for Circular Polarization Diversity," Microwave and Optical Technology Letters, vol. 50, no. 9, September 2008, pp. 2348-2350. doi:10.1002/mop.23674

12. A. Mirkamali, P. S. Hall and M. Soleimani, "Reconfigurable Printed-Dipole Antenna with Harmonic Trap for Wideband Applications," Microwave and Optical Technology Letters, vol. 48, no. 5, May 2006, pp. 927-929.

13. C. N. Zhang, S. N. Yang, H. K. Pan, A. E. Fathy, S. El-Ghazaly and V. K. Nair, "A Reconfigurable Multiband Patch Antenna for Wireless Applications Using MEMS Switches," Microwave and Optical Technology Letters, vol. 51, no. 8, August 2009, pp. 1892-1896.

14. F. Tefiku and C. A. Grimes, "Design of Broad-Band and Dual-Band Antennas Comprised of Series-Fed PrintedStrip Dipole Pairs," IEEE Transactions on Antennas and Propagation, vol. 48, no. 6, 2000, pp. 895-900.

15. A. A. Eldek, "Pattern Stability Optimization for Wideband Microstrip Antennas for Phased Arrays and Power Combiners," Microwave and Optical Technology Letters, vol. 48, no. 8, August 2006, pp. 1492-1494.

16. A. A. Eldek, "Design of Double Dipole Antenna with Enhanced Usable Bandwidth for Wideband Phased Array Applications," Progress in Electromagnetics Research, Vol. 59, 2006, pp. 1-15.

17. Ansoft Corporation, "HFSS: High Frequency Structure Simulator Based on the Finite Element Method," Version 12, 2008.

10

CHAPTER

STUDY ON MINIATURIZED UHF ANTENNAS FOR PARTIAL DISCHARGE DETECTION IN HIGH-VOLTAGE ELECTRICAL EQUIPMENT

Jingcun Liu, Guogang Zhang *, Jinlong Dong and Jianhua Wang

State Key Lab of Electrical Insulation and Power Equipment, Xi'an Jiaotong University, Xi'an 710049, China

ABSTRACT

Detecting partial discharge (PD) is an effective way to evaluate the condition of high-voltage electrical equipment insulation. The UHF detection method has attracted attention due to its high sensitivity, strong interference resistance, and ability to locate PDs. In this paper, a miniaturized equiangular spiral antenna (ESA) for UHF detection that uses a printed circuit board is proposed. I-shaped, L-shaped, and C-shaped microstrip baluns were designed to match the impedance between the ESA and coaxial cable and were verified by a vector network analyzer. For comparison, three other types of UHF antenna were also designed: A microstrip patch antenna, a microstrip slot antenna, and a printed dipole antenna. Their antenna factors were calibrated in a uniform electric field of different frequencies modulated in a gigahertz transverse electromagnetic cell. We performed comparison experiments on PD signal detection using an artificial defect model based on the international IEC 60270 standard. We also conducted time-delay test experiments on the ESA sensor to locate a PD source. It was found that the proposed ESA sensor meets PD signal detection requirements. The sensor's compact size makes it suitable for internal installation in high-voltage electrical equipment.

Keywords: equiangular spiral antenna; high-voltage electrical equipment; microstrip balun; partial discharge; ultra-high-frequency antenna

1. INTRODUCTION

The safety and reliability of high-voltage electrical equipment helps ensure the stable performance of a power system, but insulation breakdowns account for approximately 80% of total failures [1]. Therefore, it is important to assess, diagnose, and predict the insulation status of high-voltage electrical equipment. Partial discharge (PD) is low-energy ionization or a small electrical spark occurring in an insulating component that lacks permittivity and dielectric strength homogeneity. As those characteristics decline, the number and magnitude of the PD pulses increase rapidly, which may result in flashover and total breakdown. The presence of PD is a common symptom of insulation deterioration. Consequently, PD detection is a feasible way to avoid insulation degeneration. PD detection is now compulsory for judging the health and remaining life of high-voltage electrical equipment such as switchgear, transformers, and gas insulated switchgear (GIS) [2,3,4,5,6,7]. Several methods of PD detection and measurement have been developed and applied, such as high-frequency pulse current (HFPC) method; optical, chemical, and ultrasonic methods; radio frequency (RF) method; transient earth voltage (TEV) method and ultra-high frequency (UHF) method [8,9,10,11,12,13,14,15].

Among those methods, the UHF method is practical and effective due to its accuracy in locating PD sources and its convenience for online PD monitoring [16,17]. A key component of the UHF sensor is its antenna, which is important for acquiring and identifying UHF signals. UHF antennas can be either internal or external, depending on whether they are mounted inside or outside the high-voltage electrical equipment. External antennas are limited by their low sensitivity because of the shielding effect of the equipment casing and the interference or noise from the electromagnetism environment. This necessitates a close study of internal antennas. The width of a PD pulse could be as low as several nanoseconds or even sub-nanoseconds, so it is generally accepted that the UHF signal excited by PD lies in the range of 300 MHz to 3 GHz. This wide frequency range means that UHF internal antenna performance depends strongly on the antenna's dimensions.

As a result, researchers have long been interested in optimizing the antenna's broadband characteristics and minimizing antenna size. Li presented a novel compact multiband Hilbert fractal microstrip antenna for transformers, taking into account the frequency band and suitable size. The antenna was small, with a bandwidth of a few hundred MHz, but it also had very low gain [3]. Based on the designs of horn antennas, biconical log-periodic antennas, loop antennas, and dipole antennas, Kaneko built an improved dipole antenna and achieved high sensitivity [5]. Álvarez proposed optimized high-frequency current transformers (HFCT) and UHF sensors for online PD measurement [6]. Robles chose four types of antennas, including two monopoles, a zigzag antenna, and a log-periodic antenna, to measure PDs and contrasted their differing frequency behaviors. That research showed that a monopole of 5 cm was the best option [8]. Ye manufactured a multiband antenna using loop-antenna theory and a meandering technique. It worked in bandwidth ranges of 480–520, 800–850, and 1100–1200 MHz [12]. Li designed an internal two-arm equiangular spiral antenna (ESA) for gas insulated switchgear that greatly improved UHF antenna performance, but its impedance transformer made it unsuitably large [10].

To summarize, the existing literature covers UHF antennas in GIS, transformers, switchgears and XLPE cables. However, antennas with known wideband performance, such as log-periodic or spiral antennas, are always significantly large in size. For high-voltage electrical equipment, an internal UHF antenna is required for both its performance and its size. In the research for this paper, a sensor comprising a dual-arm planar ESA and a miniaturized microstrip balun was fabricated and tested. Its design was based on printed circuit board (PCB) technology. In addition, three other UHF antennas—a microstrip patch antenna (MPA), a microstrip slot antenna (MSA), and a printed dipole antenna (PDA)—were also fabricated and tested to compare them with the ESA sensor. Critical performance parameters in the UHF range were optimized through simulation and verified by a vector network analyzer (VNA). Using a standard EM field modulated in a gigahertz transverse electromagnetic (GTEM) cell, the antenna factors (AFs) of those sensors, which describe receiving characteristics to determine electric field strength, were measured and fitted. Based on the PD quantity reference provided by the traditional IEC 60270 method, and an artificial defect model as a PD source, we tested the performances of the proposed sensors and their time-delay characteristics in locating a PD source.

This paper has five parts: Section 1 is the introduction; Section 2 explains the design and optimization of ESA, its impedance balun, and the other three antennas; Section 3 describes the measurement of AFs in a GTEM cell; Section 4 shows the UHF signal detection of the antennas; and Section 5 presents the conclusions.

2. DESIGN OF THE UHF ANTENNAS

Various components must occupy the compact inner space of high-voltage electrical equipment. Because of this, internal UHF sensors should meet the following requirements: compact dimensions, strong anti-interference ability, good directivity, and high gain. A UHF sensor should work in the frequency range of 300 MHz to 3 GHz. Frequency independent antennas, such as the Archimedean spiral antenna, the ESA, and the log-periodic dipole antenna, which perform the same way at every frequency, are a perfect choice for this application. However, the disadvantage of most frequency independent antennas is that they are too large to be installed inside high-voltage electrical equipment, so much effort has been made to reduce their size. The ESA, developed in the 1950s, is an ideal type of frequency independent antenna. When requirements are not excessively restrictive, the ESA can accommodate a wide operational bandwidth while being reduced to a reasonable size. Therefore, an ESA was chosen as the design prototype in this work. MPA, MSA, and PDA were also built and tested for comparison.

In practice, the ESA often suffers from lack of an appropriate planar feeding way. A balun (balance–unbalance) element is used to change a balanced feed to an unbalanced feed; i.e., the ESA to a coaxial cable. Another function of the balun is impedance conversion. A typical balun is designed vertically, but that makes it too long to install in the limited internal space of high-voltage electrical equipment. In this paper, three kinds of modified horizontal baluns were proposed to significantly reduce the overall size of the ESA. The performance of the designed ESA with the horizontal baluns were measured and studied by a VNA.

2.1. Design of a Dual-Arm ESA

To create an ESA, in the polar coordinates system the equation of one spiral curve should be written as:

$$\rho = \rho_0 e^{a\varphi} \; \rho \leq \rho_{max} \tag{1}$$

where ρ is the polar radius, which is the radial distance between the center of the spiral and the point at the polar angle φ, such that ρ_0 is the radial distance at $\varphi = 0$, and ρ_{max} refers to the radial distance at $\varphi = \varphi_{max}$. The rate of wrapping is determined by the constant a. The equiangular spiral has the property that at any two points the angles between the tangent and radial vectors are equal. In fact, the wrap angle α can be related to a by:

$$\tan\alpha = \frac{1}{a} \tag{2}$$

For a certain frequency, when the arm length from vertex to edge is equal to wavelength λ, the corresponding radial length is approximately $\lambda 4$. The relationship between wavelength and frequency is:

$$\lambda = \frac{c}{f} \tag{3}$$

where c is the speed of light. Therefore, the inner and outer radii of the ESA can be calculated for the bandwidth of interest. For example, ρ_0 is defined by the superior frequency limit:

$$\rho_0 = \frac{\lambda_{min}}{4} \tag{4}$$

The radial distance ρ_0 is 2.5 mm because the working band is up to 3 GHz. The outer radius ρ_{max} is defined by the inferior frequency limit:

$$\rho_{max} = \frac{\lambda_{max}}{4} \tag{5}$$

Because the minimum working frequency is 500 MHz, ρ_{max} should be 150 mm, but this dimension was reduced to 62 mm to meet the installation requirements of the sensor. In other words, low-frequency performance was sacrificed to its practical application. With the above curve drawn, rotating it repeatedly by 90° formed the edges of two arms of the ESA, which made the ESA a self-complementary structure (Figure 1).

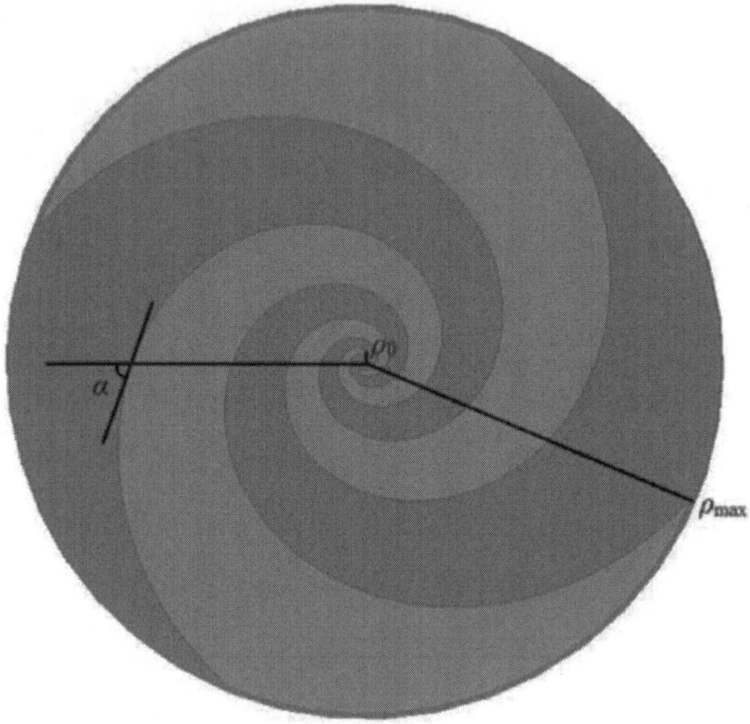

Figure 1. Geometry of a dual-arm ESA.

Other determined parameters of ESA are listed in Table 1.

Table 1. Parameters of the designed ESA.

Wrap Angle	Rotation Angle	Substrate Thickness	Relative Permittivity
70°	3 π	1.6 mm	4.4

In practice, the impedance was approximately 140 Ω, according to simulation and measurement [16,17]. The antenna was connected to a 50 Ω coaxial cable, so an exponential asymptote microstrip balun was designed to provide broadband matching. Figure 2 shows the structure of the balun. Its total length was 220 mm. The width at the feed point was 0.5 mm; the width at the other end was 4.5 mm.

Figure 2. Structure of the exponential asymptote microstrip balun.

Initially, the microstrip balun was connected to ESA vertically, as shown in Figure 3.

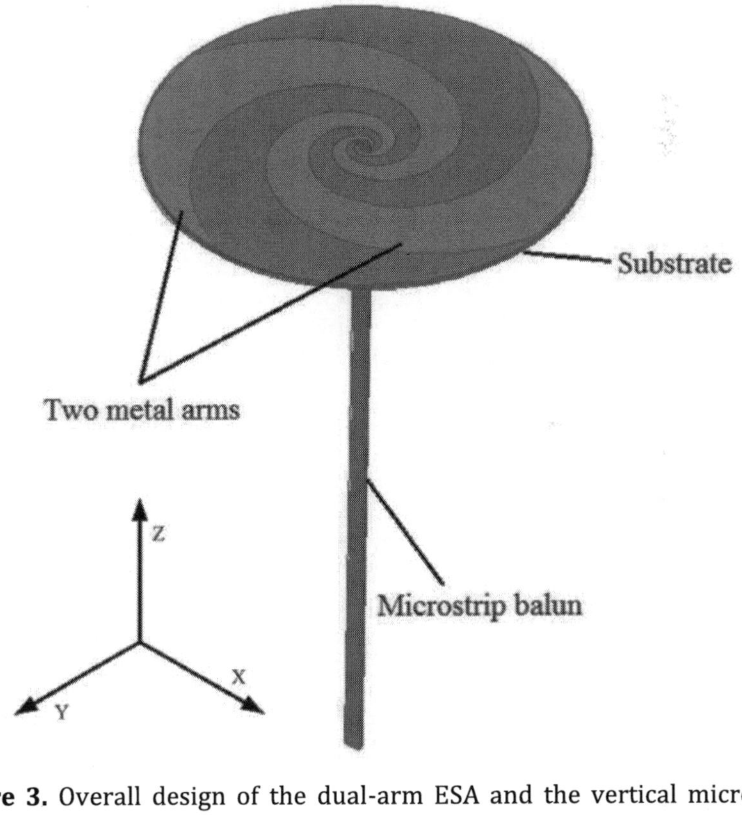

Figure 3. Overall design of the dual-arm ESA and the vertical microstrip balun.

We conducted a thorough simulation using the three-dimensional EM simulation software ANSYS HFSS. A linear frequency sweep of 500 MHz to 3 GHz and a 10-MHz step size was configured in the simulation. The results of the key parameters of the designed ESA are shown in Figure 4 through Figure 6.

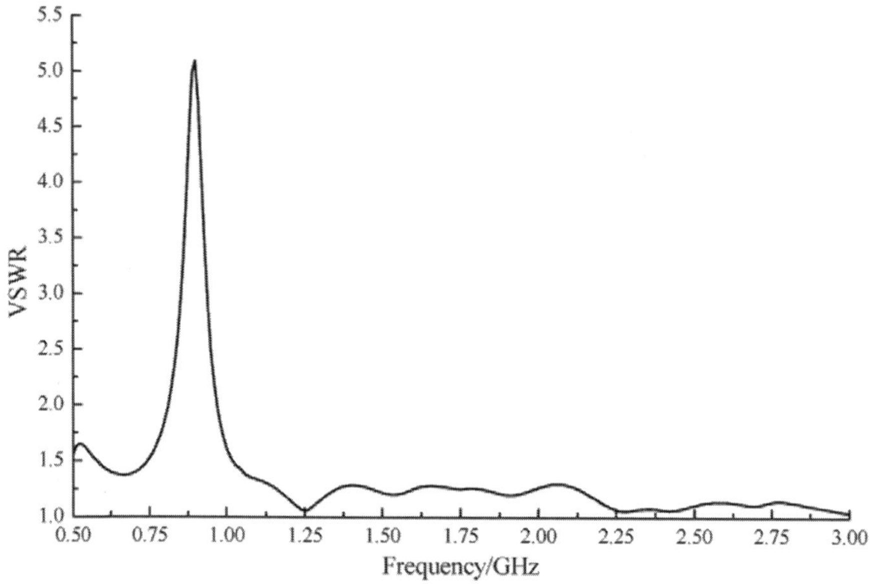

Figure 4. Voltage standing-wave ratio (VSWR) simulation results of the experimental ESA.

The VSWR indicates the impedance match between an antenna and a feed line. The smaller the VSWR is, the better the match. In normal applications, an antenna works well in a frequency band where the VSWR is less than 2. In Figure 4, the frequency band whose VSWR is less than 2 lay almost in the whole sweep range, except for the 820 MHz to 960 MHz portion.

The VSWRs of the other frequencies were close to 1.25 and remained stable, which indicated little reflection loss and decent antenna performance. The working bandwidth was 2.36 GHz and the relative bandwidth was up to 94.4%. It meant that even though low-frequency performances were sacrificed to reduce the size of the designed ESA, that antenna could still meet the signal detecting requirement.

Input impedance is the ratio of the voltage to the current at an antenna input [18]. Because 50 Ω coaxial cable is widely used, it is required that the real input impedance approach 50 Ω and the imaginary impedance be as near to 0 as possible.

Figure 5 shows impedance matching conditions in our tests. The real input impedance fluctuated around 50 Ω and the imaginary input impedance fluctuated around 0. The scope of the fluctuation under 1 GHz was approximately 35 Ω; above 1 GHz the scope was within 10 Ω. Thus, the exponential asymptote microstrip balun did provide wideband impedance matching for the ESA. Gain describes the ratio of the power density of an actual antenna to an ideal radiating element in the same position under the same conditions. Realized gain is the term used when reflection loss due to impedance is taken into account. It is the ratio of power radiated to the power input of the antenna. Generally speaking, the higher the realized gain, the stronger the signal a UHF antenna receives. The directivity of an antenna refers to the distribution of the energy it radiates [19]. A radiation pattern is used to judge the directivity of an antenna in practical applications. Figure 6 shows the radiation pattern of the XZ and YZ planes at a center frequency of 1.75 GHz. At marker 1, the peak realized gain was up to 3.24 dB. The results also show the irregularity of the ESA's directivity, which means it is a unidirectional antenna. As a result, the angle and position of the ESA's installation must be considered in experimental and practical applications.

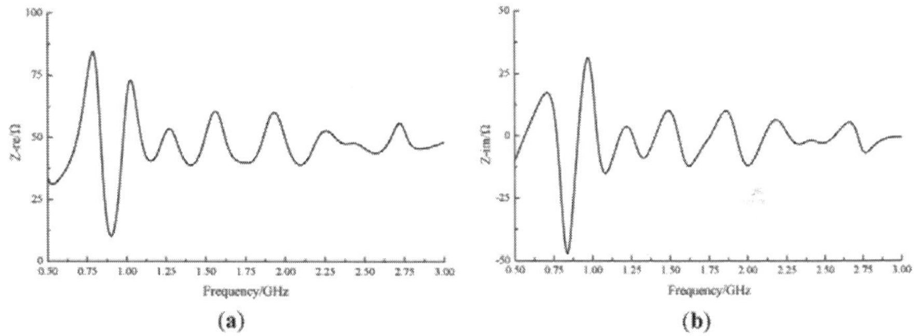

Figure 5. Impedance simulation results of the ESA: (**a**) real impedance; (**b**) imaginary impedance.

2.2. Design and Optimization of the Balun

Clearly, an ESA with a long vertical balun performs well but is oversized. To reduce the size of the designed sensor for an internal installation, three main approaches are presented in the following. The prototype ESA is shown in Figure 7a, and three types of microstrip baluns are shown in Figure 7b through Figure 7d.

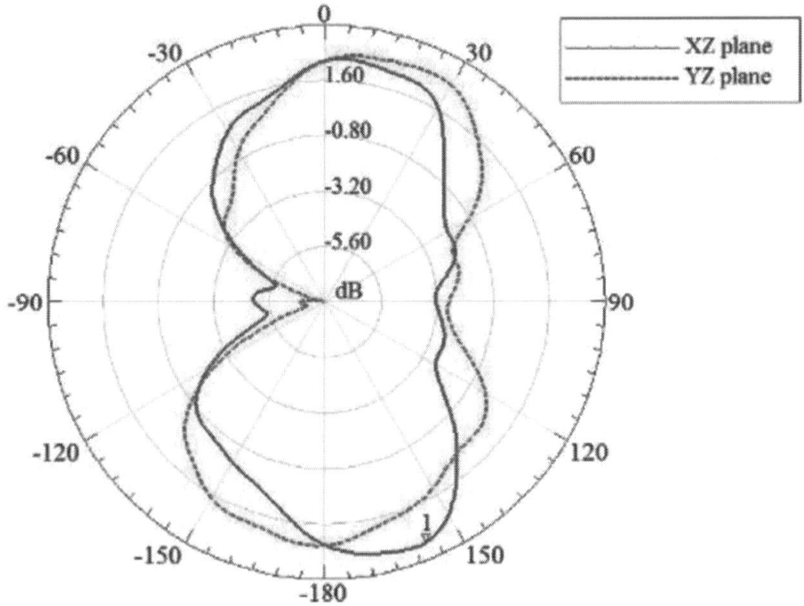

Figure 6. Radiation pattern simulation results of the ESA.

(1) Horizontal Balun

As discussed above, a vertical balun adds unwanted depth and hinders reliable installation. The first improvement approach was to simply place the same exponential asymptote balun, called an I-shaped microstrip balun, horizontally, as shown in Figure 7b. This slightly increased the length of the whole sensor, but clearly reduced the 3D space the sensor occupied.

(2) Deformed Balun

With the metal ground plate patch on one side of the substrate remaining unchanged, the second approach deformed the shape of the microstrip balun on the other side of the substrate. As shown in Figure 7c, the L-shaped microstrip balun was designed by bending the asymptote line 90°. Figure 7d shows an innovative spiral balun called a C-shaped microstrip balun. The two edges of the line were based on an Archimedean spiral, with equal widths at the feed point and the other end. The L-shaped and C-shaped baluns were shorter than the original balun.

(3) Integrated Balun

Ideas to integrate feed into antennas are provided in previous studies. Constructions similar to the I-shaped and C-shaped baluns have been designed and tested [20,21,22,23]. Integrating baluns into a spiral antenna renders the entire system two-dimensional. Results show that antennas fed by this method work well in the ultra-wideband (UWB) range of 3.1 GHz to 10.6 GHz. However, for the UHF band an integrated balun could occupy more space and be difficult to manufacture. Therefore, in this study that method is not considered.

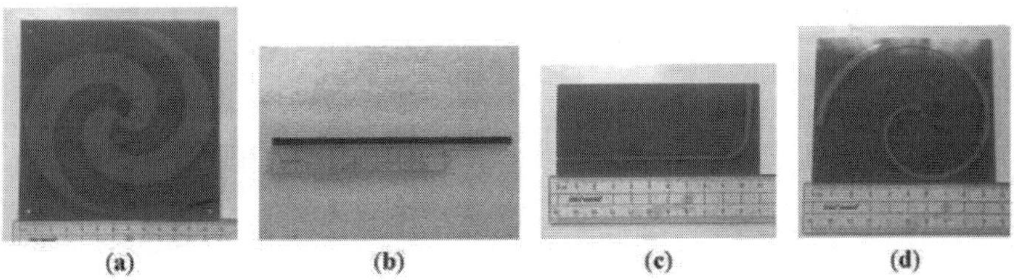

Figure 7. Manufactured prototype dimensions of the equiangular spiral antenna (ESA) and microstrip baluns. (**a**) ESA; (**b**) I-shaped balun; (**c**) L-shaped balun; (**d**) C-shaped balun.

Figure 8 indicates how the ESA and baluns connect to the back of the antenna. Obviously, a sensor made of the ESA and a vertical balun occupies too much space. The modified baluns can be placed on almost the same plane as the ESA, which is desirable for installation and fixation.

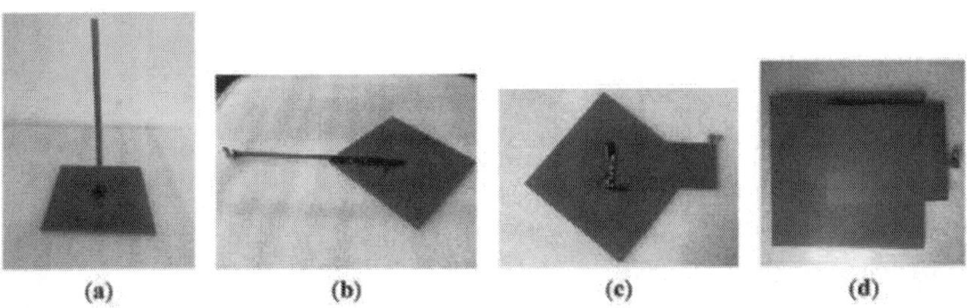

Figure 8. Connections of the ESA and baluns. (**a**) vertically; (**b**) with an I-shaped balun; (**c**) with an L-shaped balun; (**d**) with a C-shaped balun.

To find the best of the optimized baluns, simulations and measurements were conducted. The simulation configuration was also a linear sweep of 500 MHz to 3 GHz with a step size of 10 MHz. VSWRs of four antennas were measured by an Agilent E5061B VNA. Figure 9 shows the comparison results of the four types of baluns. Under 1 GHz, simulation results were very similar to those of the real performance. Over 1 GHz, measured VSWRs were higher due to the limitations of the VNA and the high-frequency loss of coaxial cable. In general, broadband performance of the three optimized baluns was similar to that of the original vertical balun; the I-shaped balun was the closest. For the original vertical balun, the working band where the VSWR was less than 2 was 500–600 MHz and 800 MHz–3 GHz, and bandwidth was 2.25 GHz, which suggests that deviation from simulation was less than 6%. Meanwhile, the I-shaped balun performed slightly worse but much like the vertical type. Its working band included 500–550 MHz, 850–1050 MHz, and 1.15–3 GHz; bandwidth was 2.1 GHz and relative bandwidth was 84%. The VSWR of the L-shaped and C-shaped baluns at low frequencies increased substantially. Therefore, the ESA with the I-shaped microstrip balun was chosen for the following experiments.

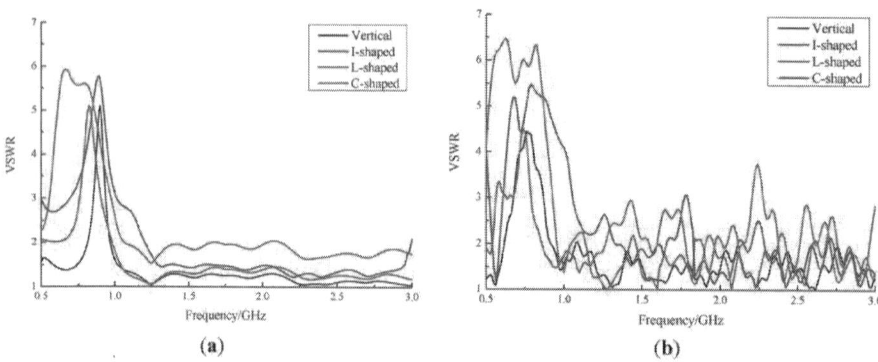

Figure 9. Comparison results of four types of baluns: (**a**) simulation; (**b**) VNA measurement.

2.3. Design of Other Antennas

In the field of communications, many types of antennas that work in the UHF band, such as the microstrip antenna, have been used. Previous studies on UHF detection of PD have long been focused on these common types of antennas. Their main advantages are their small size, design simplicity, robustness, and convenient adaptation to mass production using PCB technology. Therefore, in this paper three miniaturized antennas are designed and printed on dielectric substrates to compare with the ESA and determine the optimal antenna for PD measurement.

Figure 10a illustrates the structure of an MPA that has been widely used in recent years. It is fabricated by etching the antenna element pattern on a metal trace. The trace is bonded to an insulating dielectric substrate, with a continuous metal layer bonded to the opposite side of the substrate to form a ground plane. For impedance matching with a coaxial cable, a one-quarter wavelength impedance transformer is added. Similarly to an MPA, as shown in Figure 10b, an MSA is made by slotting on its metal ground plate. In the illustrated design, a feed and meandering method have been used to aid miniaturization. A PDA is a deformation of a half-wave dipole antenna. The version shown in Figure 10c has a metal arm on each side of the dielectric substrate. Because it is a balanced structure, this PDA must incorporate a triangular balun to balance the antenna with the coaxial cable.

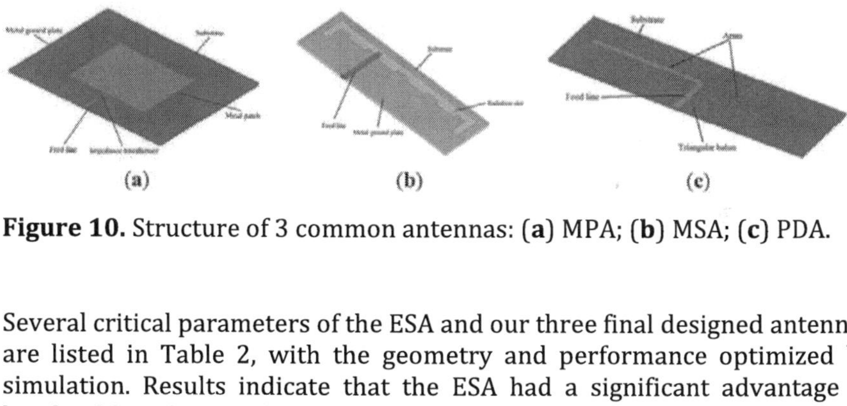

Figure 10. Structure of 3 common antennas: (**a**) MPA; (**b**) MSA; (**c**) PDA.

Several critical parameters of the ESA and our three final designed antennas are listed in Table 2, with the geometry and performance optimized by simulation. Results indicate that the ESA had a significant advantage in bandwidth and gain, while the MSA and PDA, with their small geometry and good directivity, are easy to install. Overall, each of the four antennas meets the requirements of PD detection.

Table 2. Parameters of four types of UHF antennas.

Type of Antenna	Geometry in mm	Working Band in MHz	Directivity	Peak Realized Gain/dB
ESA	124 × 124 × 1.6	500–550 850–1050 1150–3000	Poor	3.24
MPA	188 × 138 × 1.6	920–1020	Poor	−1.54
MSA	100 × 30 × 1.6	961–979	H plane good	−0.17
PDA	140 × 40 × 1.6	908–1029	H plane omnidirectional	1.40

3. CALIBRATION OF THE ANTENNA FACTOR

To obtain a quantitative relationship between measured electric field strength and antenna output voltage, antenna calibration is required. The antenna factor is defined mathematically as:

$$AF = \frac{E_{incident}}{V_{received}} \quad (6)$$

It measures the received voltage in the presence of an electric field. For PD detection applications, once the AF of an antenna is determined, electric field strength on the reference plane excited by a PD source can be calculated. That helps to judge the insulation condition of the target equipment. In this study, antenna calibration took place in the standard electric field generated inside a GTEM cell. That device is a large, tapered coaxial structure, with transverse electromagnetic waves excited around the core board. A GTEM cell provides an EM field with great uniformity and reproducibility in a desired region of space. Using the uniform electric field within a workspace under different frequencies, we conducted experiments in a GTEM cell to measure AFs.

3.1. Experimental Setup

The experiment configuration is shown in Figure 11. The signal generator, connected to the GTEM cell as excitation source, modulated the needed frequencies and field strengths of the EM field. The four types of antennas were placed in the center zone of the test cell (one-third of the septum height). In that area of the GTEM cell, the electric field vector is vertical and the EM wave should be considered as linearly polarized. Therefore, to achieve the maximum received signal and the least loss from polarization mismatch, the electric field vectors of the antennas should match. For horizontally linearly polarized antennas like MPAs, MSAs, and PDAs, the E plane is positioned in parallel with the vertical electric field vector. For circularly polarized antennas like ESAs, the orientation could be flexible as long as the antenna axis is horizontal, so during the experiment, all four antennas were oriented vertically. Their received signals were output to an R&S ESR EMI test receiver through a 50 Ω coaxial cable. Before the signal generator was turned on, a modulation range of frequencies was set. During the experiment, the output signals of UHF antennas were recorded by the EMI receiver at every frequency.

Figure 11. Experimental setup for calibration of AFs.

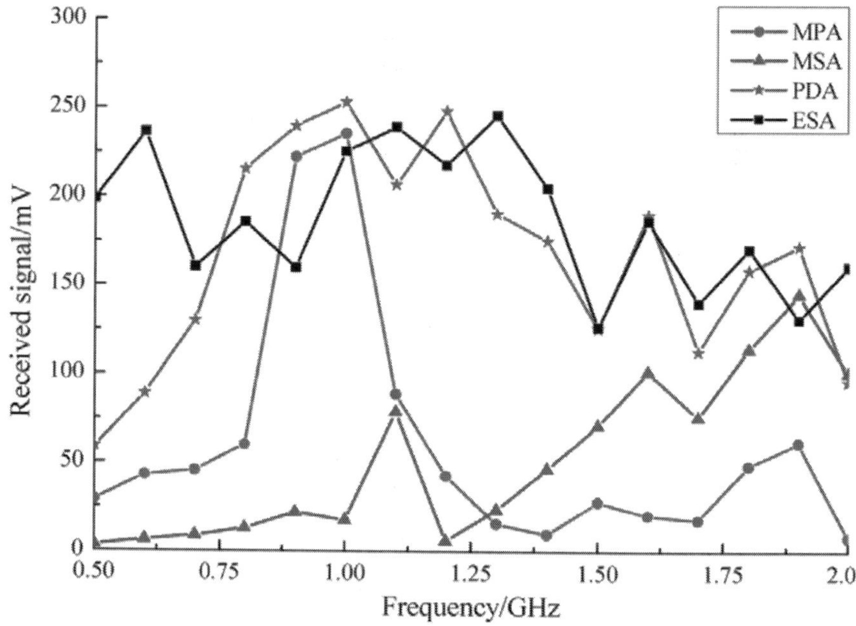

Figure 12. Frequency sweep results from testing in the GTEM cell.

3.2. Experimental Result and Antenna Factor Determination

Figure 12 shows the frequency sweep (500 MHz to 2 GHz, step size 100 MHz) results of the ESA and three other sensor prototypes. Results suggest that the amplitude of the signal the ESA detected was over 150 mV and remained stable in the sweep range. For the PDA, the amplitude of the output signal at frequencies below 700 MHz was low, while beyond 700 MHz the amplitude increased to almost the same as that of the ESA. For the MPA, the signal amplitude was highest near the 970 MHz center frequency, but lower than 50 mV at other frequencies. For the MSA, the signal amplitude beyond 1 GHz grew higher, while it was very low at lower frequencies.

Antenna factor (AF) expresses the relationship between an output signal and electric field strength, which is a function of frequency (f, GHz). We based our calibration of AFs on previous experimental data. Five times polynomial fitting was done by delineating a scatterplot of calculated AF values. Table 3 shows the coefficients of fitting Equation (7):

$$AF = B_5 f^5 + B_4 f^4 + B_3 f^3 + B_2 f^2 + B_1 f + C \tag{7}$$

Table 3. Coefficients of the fitting equation of AFs.

Type of Sensor	*C*	B_1	B_2	B_3	B_4	B_5
MPA	−2781.3	21,840	−53,003	55,908	−26,426	4602.6
MSA	30,694	−119,108	182,850	−135,273	48,086	−6592.8
PDA	878.7	−2298.6	2033.5	−475.5	166.9	67.0
ESA	−663.7	3427.1	−6121.1	5132.7	−2040.7	311.2

As shown in Figure 13, the AF of the proposed ESA sensor was the lowest and stabilizes at approximately 50 in the whole band, while the AFs of others were higher and fluctuated. The ESA was the most sensitive antenna.

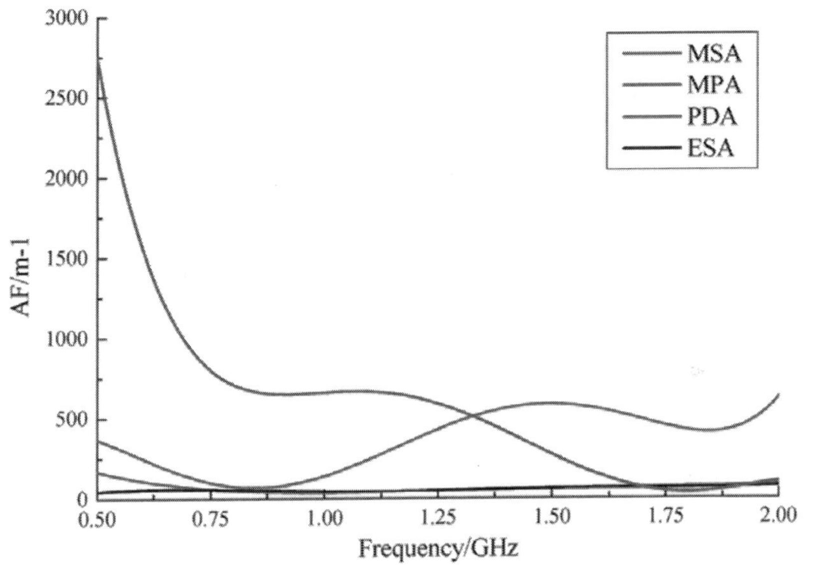

Figure 13. Calibration curves of antenna factors for four antenna types.

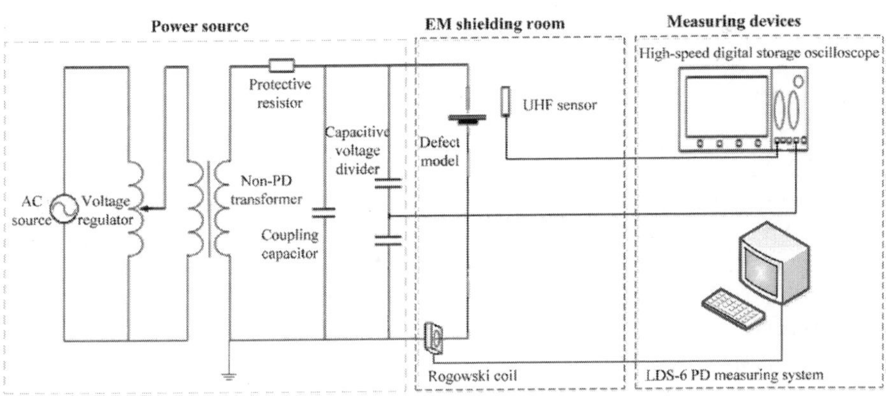

Figure 14. The PD experiment platform.

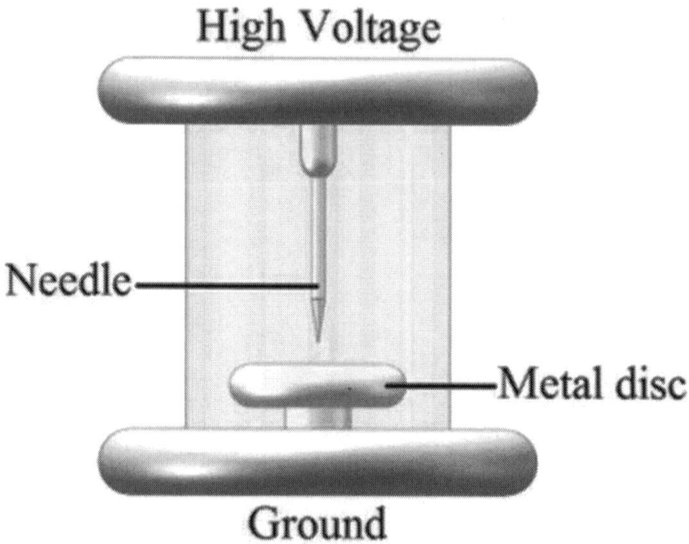

Figure 15. The artificial defect model.

4. UHF DETECTION EXPERIMENTS

To study the performances of the antennas developed in this paper, a PD experiment platform was set up in an EM shielding room, as illustrated in Figure 14. The high-voltage power source of the circuit was provided by a non-PD testing transformer. To simulate a PD source, we designed a needle-plane electrode system (Figure 15) as a typical artificial defect model. An LDS-6 PD measuring system was adopted to measure the PD magnitude, which was based on the IEC 60270 standard and had a resolution of 1 pC. A Tektronix DPO7104 high-speed digital storage oscilloscope was used as the signal recorder, with an input channel bandwidth of 1 GHz, a sampling rate of 20 GHz, and a memory depth of 32 M.

4.1. Comparison Experiment of PD Detection for the Four Sensors

In the experiment, the four antennas were placed 2 m from the defect model and connected to an oscilloscope through coaxial cable to display the output signal. When the discharge quantity on the LDS-6 system stabilized, the UHF signal sample of each antenna on the oscilloscope was recorded.

The reference discharge quantity was 320 pC; the collected PD UHF single-cycle waveforms of the four sensors are shown in Figure 16. The signal magnitude of the ESA was up to approximately 35 mV and the waveform detail was clear. Moreover, the measured sensitivity of the ESA reached 16 pC when the background noise was 8 pC. The signal magnitude of the MPA was approximately 4.5 mV, so it was hard to detect low-energy PD or PD when background noise was high. The signal magnitude of the MSA was approximately 8 mV, but its waveform was difficult to restore effectively due to its narrow bandwidth. For the PDA, the signal magnitude was approximately 15 mV, but the symmetry of its signal was poor, because the negative part of the waveform was not as clear as the positive part.

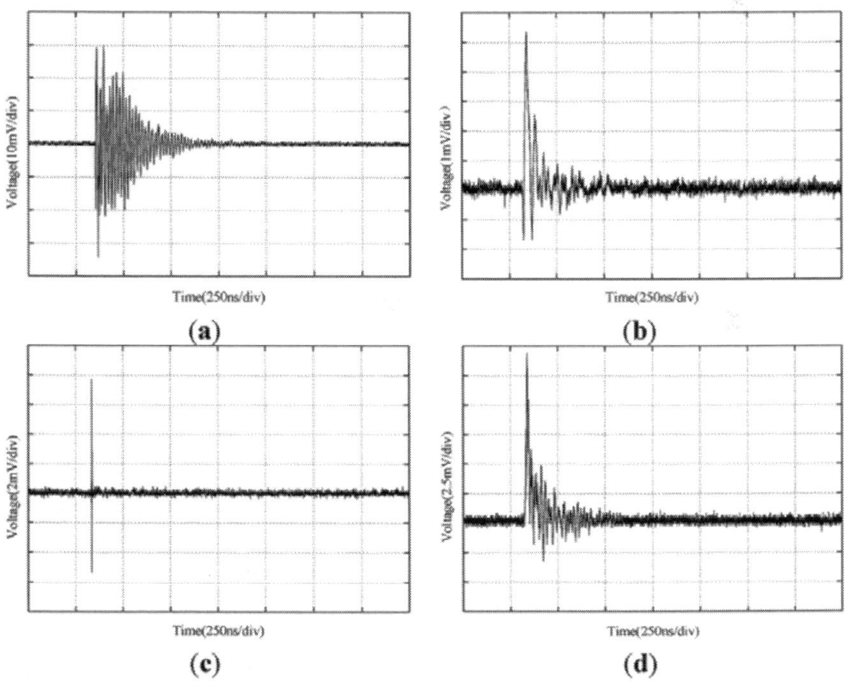

Figure 16. UHF waveforms of the PD comparison experiment: (**a**) ESA; (**b**) MPA; (**c**) MSA; (**d**) PDA.

4.2. Time-Delay Experiment

A triangulation method is widely used to detect PD locations [24,25,26]. By measuring the delay time of a UHF pulse received by two sensors in different locations, the distance between the two sensors can be calculated by:

$$L = ct \tag{8}$$

where c is the speed of light and t is the measured delay time. In practical applications it usually takes four sensors to determine a PD source location in 3D space using the following equations:

$$\begin{cases} (x-x_1)^2+(y-y_1)^2+(z-z_1)^2=[cT]^2 \\ (x-x_2)^2+(y-y_2)^2+(z-z_2)^2=[c(T+t_{12})]^2 \\ (x-x_3)^2+(y-y_3)^2+(z-z_3)^2=[c(T+t_{13})]^2 \\ (x-x_4)^2+(y-y_4)^2+(z-z_4)^2=[c(T+t_{14})]^2 \end{cases} \tag{9}$$

where (x, y, z) are the coordinates of the source, (x_i, y_i, z_i) are the coordinates of sensor i, T is the time from source to the first sensor, and t_{1i} is the delay time of sensor i and the first sensor.

To determine a PD source location in 2D space, we placed 3 ESA sensors at different locations and connected them to three channels of the oscilloscope by coaxial cables of identical lengths. Figure 17 shows the positions of the three ESA sensors and the artificial defect model. To prove the feasibility and accuracy of this method using the proposed antenna, the PD source position was calculated by Equation (9) and compared with the preset position. The collected PD UHF waveforms of the three sensors are shown in Figure 18. It can be seen that the delay time between Antennas 1 and 2 was 6 ns and between Antennas 1 and 3 was 11 ns. Thus, the calculated PD source coordinate was approximately (−0.05, 0.01), which means the deviation from the actual source location was only 5 cm or less.

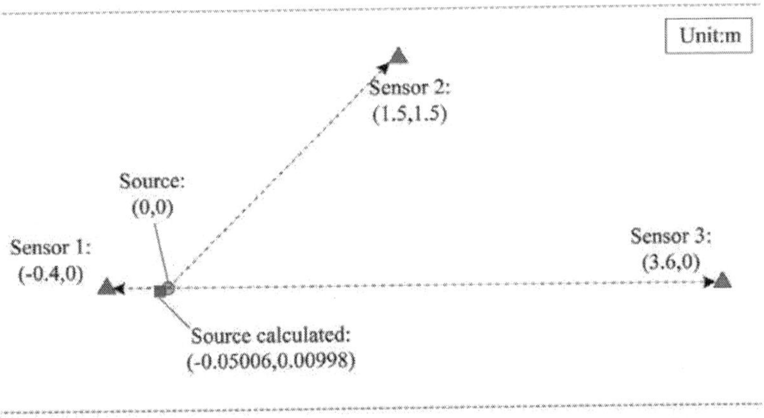

Figure 17. Schematic diagram of the positions of the source and the three sensors.

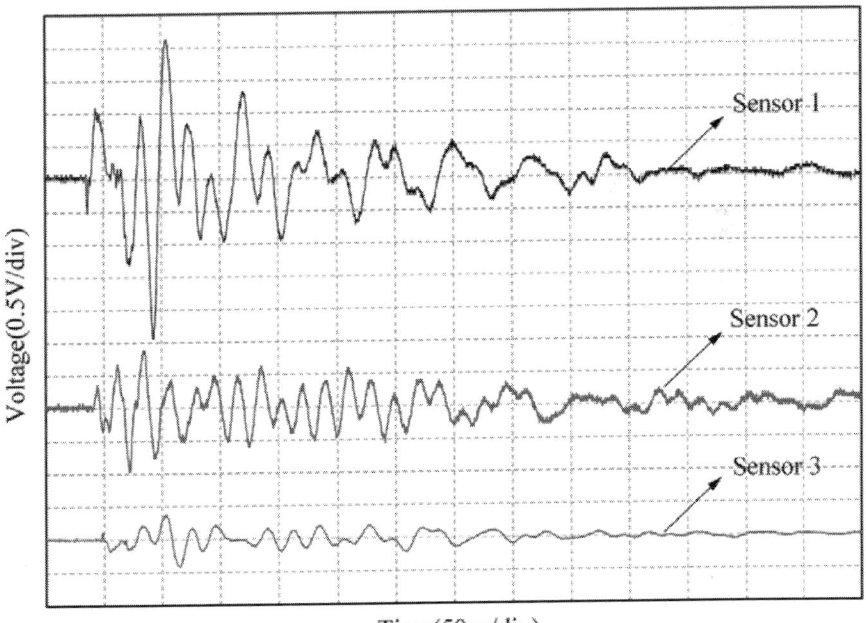

Figure 18. Time-delay experiment waveforms of three ESA sensors.

5. CONCLUSIONS

This study focused on the design and optimization of internal antennas for UHF detection of PDs in high-voltage electrical equipment. The proposed dual-arm ESA satisfied the requirements of both broadband performance and compact size. An MPA, an MSA, and a PDA were also developed and tested for comparison. A GTEM cell was used for calibrating AFs. A PD detecting experiment platform was set up to test the performances of the proposed antennas. A time-delay experiment to locate the PD source was also conducted. The results of the study can be summarized as follows:

(1) The proposed antenna was designed using the basic configuration of an ESA, while its outer radius was reduced to 62 mm to meet the strict installation requirements of an internal sensor. The key parameters were determined through simulation. It was found that an ESA can work in the UHF band range with high realized gain and good impedance matching, although its directivity is not ideal and should be considered.

(2) An I-shaped balun, an L-shaped balun, and a C-shaped balun are attempted to further reduce the size of the entire sensor. A VSWR measurement on a VNA was conducted. The I-shaped horizontal balun performed like the original vertical type with a wide working band (500–550 MHz, 850–1050 MHz, and 1.15–3 GHz); the 2.1 GHz bandwidth was the best option.

(3) An MPA, an MSA, and a PDA were designed, and their performances were also analyzed and optimized. The main differences between the ESA and the other three types of sensors lay in bandwidth and the AF. The result of calibration in the GTEM cell indicated that the AF of the proposed ESA sensor was the lowest and most stable, while the AFs of the others were higher and fluctuated.

(4) The experiment on UHF detection of PD by the four types of sensors showed that the performance of the ESA sensor surpassed those of the others. The magnitude of signal received by the ESA sensor was the largest, and the shape and details of the PD waveform were clearly visible on the oscilloscope. An experiment using a triangulation method to locate a PD source using the proposed ESA was also conducted. It was found that the proposed ESA revealed a PD source location with a precision of 5 cm.

ACKNOWLEDGMENTS

This work was supported by the National Natural Science Foundation of China (51477129).

AUTHOR CONTRIBUTIONS

Jingcun Liu designed the antenna models and carried out the analysis of their application in PD measurements. He also conceived and performed the detection experiments with Jinlong Dong. Guogang Zhang proposed the research topic, offered valuable suggestions and interesting discussions in all research activities. He and Jianhua Wang also provided theoretical guidance and supported the study. Jingcun Liu and Guogang Zhang wrote and organized the manuscript.

CONFLICTS OF INTEREST

The authors declare no conflicts of interest.

REFERENCES

1. Sarkar, B.; Mishra, D.K.; Koley, C.; Roy, N.K. Microstrip patch antenna based UHF sensor for detection of partial discharge in high voltage electrical equipments. In Proceedings of the Annual IEEE India Conference, Pune, India, 11–13 December 2014.

2. Yao, C.; Chen, P.; Huang, C.; Chen, Y.; Qiao, P. Study on the application of an ultra-high-frequency fractal antenna to partial discharge detection in Switchgears. Sensors **2013**, 13, 17362–17378.

3. Li, J.; Jiang, T.; Cheng, C.; Wang, C. Hilbert fractal antenna for UHF detection of partial discharges in transformers. IEEE Trans. Dielectr. Electr. Insul. **2013**, 20, 2017–2025.

4. Sarathi, R.; Koperundevi, G. Investigation of partial discharge activity of single conducting particle in transformer oil under DC voltages using UHF technique. IET Sci. Meas. Technol. **2009**, 3, 325–333.

5. Kaneko, S.; Okabe, S.; Yosh, M.; Muto, H.; Nishida, C.; Kamei, M. Detecting characteristics of various type antennas on partial discharge electromagnetic wave radiating through insulating spacer in gas

insulated switchgear. IEEE Trans. Dielectr. Electr. Insul. **2009**, 16, 1462–1472.

6. Álvarez, F.; Garnacho, F.; Ortego, J.; Sánchez-Urán, M.Á. Application of HFCT and UHF sensors in on-line partial discharge measurements for insulation diagnosis of high voltage equipment. Sensors **2015**, 15, 7360–7387.

7. Wang, L.; Fang, N.; Wu, C.; Qin, H.; Huang, Z. A fiber optic PD sensor using a balanced Sagnac interferometer and an EDFA-Based DOP Tunable fiber ring laser. Sensors **2014**, 14, 8398–8422.

8. Robles, G.; Sanchez-Fernandez, M.; Albarracin, S.R.; Rojas-Moreno, M.V.; Rajo-Iglesias, E.; Martinez-Tarifa, J.M. Antennas parameterization for the detection of partial discharges. IEEE Trans. Instrum. Meas. **2013**, 62, 932–941.

9. Hoshino, T.; Maruyama, S.; Ohtsuka, S.; Hikita, M.; Wada, J.; Okabe, S. Sensitivity comparison of disc- and loop-type sensors using the UHF method to detect partial discharges in GIS. IEEE Trans. Dielectr. Electr. Insul. **2012**, 19, 910–916.

10. Li, T.; Rong, M.; Zheng, C.; Wang, X. Development simulation and experiment study on UHF partial discharge sensor in GIS. IEEE Trans. Dielectr. Electr. Insul. **2012**, 19, 1421–1430.

11. Piccin, R.; Mor, A.; Morshuis, P.; Girodet, A.; Smit, J. Partial discharge analysis of gas insulated systems at high voltage AC and DC. IEEE Trans. Dielectr. Electr. Insul. **2015**, 22, 218–228.

12. Hai-feng, Y.; Xiu-chen, J.; Yong, Q.; Ge-hao, S.; Yue, D. Development of multi-band ultra-high-frequency sensor for partial discharge monitoring based on the meandering technique. IET Sci. Meas. Technol. **2014**, 8, 327–335.

13. Muslim, J.; Susilo, A.; Nishigouchi, K.; Kozako, M.; Hikita, M.; Arief, Y.Z.; Khayam, U.; Suwarno. Enhanced bowtie UHF antenna for detecting partial discharge in gas insulated substation. In Proceedings of the 48th International Universities' Power Engineering Conference, Dublin, Ireland, 2–5 September 2013.

14. Lopez-Roldan, J.; Tang, T.; Gaskin, M. Optimisation of a sensor for onsite detection of partial discharges in power transformers by the UHF method. IEEE Trans. Dielectr. Electr. Insul. **2008**, 15, 1634–1639.

15. Tenbohlen, S.; Denissov, D.; Hoek, S.; Markalous, S.M. Partial discharge measurement in the ultra high frequency (UHF) range. IEEE Trans.

Dielectr. Electr. Insul. **2008**, 15, 1544–1552. [Google Scholar] [CrossRef]

16. Gao, W.; Ding, D.; Liu, W.; Huang, X. Analysis of the intrinsic characteristics of the partial discharge induced by typical defects in GIS. IEEE Trans. Dielectr. Electr. Insul. **2013**, 20, 782–790.
17. Pinpart, T.; Judd, M.D. Experimental comparison of UHF sensor types for PD location applications. In Proceedings of the IEEE Electrical Insulation Conference, Montreal, QC, Canada, 31 May 2009–3 June 2009.
18. McFadden, M.; Scott, W.R. Analysis of the equiangular spiral antenna on a dielectric substrate. IEEE Trans. Antennas Propag. **2007**, 55, 3163–3171.
19. Fu, W.; Lopez, E.R.; Rowe, W.S.T.; Ghorbani, K. A planar dual-arm equiangular spiral antenna. IEEE Trans. Antennas Propag. **2010**, 58, 1775–1779.
20. Eubanks, T.W.; Chang, K. A compact parallel-plane perpendicular-current feed for a modified equiangular spiral antenna. IEEE Trans. Antennas Propag. **2010**, 58, 2193–2202.
21. Mao, S.-G.; Yeh, J.-C.; Chen, S.-L. Ultrawideband circularly polarized spiral antenna using integrated balun with application to time-domain target detection. IEEE Trans. Antennas Propag. **2009**, 57, 1914–1920.
22. Chen, T.-K.; Huff, G.H. Stripline-fed Archimedean spiral antenna. IEEE Antennas Wirel. Propag. Lett. **2011**, 10, 346–349.
23. Veysi, M.; Kamyab, M. Bandwidth enhancement of low-profile PEC-backed equiangular spiral antennas incorporating metallic posts. IEEE Trans. Antennas Propag. **2011**, 59, 4315–4318.
24. Hou, H.; Sheng, G.; Jiang, X. Robust time delay estimation method for locating UHF signals of partial discharge in substation. IEEE Trans. Power Deliv. **2013**, 28, 1960–1968.
25. Markalous, S.; Tenbohlen, S.; Feser, K. Detection and location of partial discharges in power transformers using acoustic and electromagnetic signals. IEEE Trans. Dielectr. Electr. Insul. **2008**, 15, 1576–1583.
26. Sinaga, H.H.; Phung, B.T.; Blackburn, T.R. Partial discharge localization in transformers using UHF detection method. IEEE Trans. Dielectr. Electr. Insul. **2012**, 19, 1891–1900.

11
CHAPTER
DOUBLE U-SHAPED SLOTS LOADED STACKED PATCH ANTENNA FOR MULTIBAND OPERATION

Nagendra Prasad Yadav

School of Electronic Engineering and Optoelectronic Technology, Nanjing University of Science and Technology, Nanjing, China

ABSTRACT

The design of a seven-band stacked patch antenna for the C, X and Ku band is presented. The antenna consists of an H-slot loaded fed patch, stacked with dual U-slot loaded rectangular patch to generate the seven frequency bands. The total size of the antenna is 39.25 × 29.25 mm^2. The multiband stacked patch antenna is studied and designed using IE3D simulator. For verification of simulation results, the antenna is analyzed by circuit theory concept. The simulated return loss, radiation pattern and gain are presented. Simulated results show that the antenna can be designed to cover the frequency bands from (4.24 GHz to 4.50 GHz, 5.02 GHz to 5.25 GHz) in C-band application, (7.84 GHz to 8.23 GHz) in X-band and (12.16 GHz to 12.35 GHz, 14.25 GHz to 14.76 GHz, 15.25 GHz to 15.51 GHz, 17.52 GHz to 17.86 GHz) in Ku band applications. The bandwidths of each band of the proposed antenna are 5.9%, 4.5%, 4.83%, 2.36%, 3.53%, 1.68% and 1.91%. Similarly the gains of the proposed band are 2.80 dBi, 4.39 dBi, 4.54 dBi, 10.26 dBi, 8.36 dBi and 9.91 dBi, respectively.

Keywords: H-Shape Fed Patch, Microstrip Patch Antenna, Double U-Shape Slot Loaded Stacked Patch, Multiband Antenna

1. INTRODUCTION

Antenna is a very important component of communication system. The enormous growth of mobile and satellite communication systems along with wonderful use of radars opens a huge demand to new kind of antennas such as small antennas, multi frequency antennas, and broadband antennas [1] -[3] . Planar multi-resonators and stacked microstrip patch antenna techniques are combined to yield a wide bandwidth and higher gain. Several two-layered configurations have been discussed in [4] - [6] , in which only a single patch on the bottom layer is fed by a coaxial probe and second patch on the top layer and both are electromagnetically coupled to each other. Multi-band or reconfigurable antennas are suitable candidates for providing multi-functionality. This will result in significant reductions in antenna size and cost. The primary advantage of the proposed multiband antenna lies in its ability to support two separate applications at two different frequency bands with distinctly different radiation patterns, gain and polarization characteristics using a single radiating aperture [7] . Although multi-band antennas have the capability of serving multiple frequency bands with one antenna, they are considered a weak choice in comparison to reconfigurable antennas because of crosstalk from the neighbor bands. In [8] , we have to see that the multiband antenna can be obtained by using of slot couple technique in multiple patches. In 2012, M. A. Motin, et al. [9] , presented the multiband microstrip patch antenna for X, K and Ku band application. In 2014, Jianxing Li et al. [10] , represented the multiband probe-fed stacked patch antenna for GNSS application.

In the present paper, a multiband microstrip stacked patch antenna is introduced. The proposed structure is a planar structure having all the dimensions in mm. The proposed antenna is designed by using a substrate of FR4 having thickness 1.6 mm. This proposed antenna is having seven bands of operation. The entire investigation is based on equivalent circuit model. In [11] , it represents the multi-band antenna by using stacked patch with multi- slot loaded patch, which can be compared to the proposed design. The co-axial feed technique is used for the analysis of this antenna because it occupies less space and has low spurious radiation by using Teflon connector. The proposed antenna design can be used in mobile communication, radar and satellite communication. Details of the antenna are given in the next stage.

2. ANTENNA DESIGN AND THEORETICAL CONSIDERATIONS

The design of the proposed multiband antenna is depicted in Figure 1. The upper patch is double U-shaped slot loaded stacked patch and the lower patch is the H-slot loaded fed patch. Due to the presence of the stacked patch antenna, there are two resonances associated with the two resonators [12] .

Microstrip patch with a dielectric cover is considered as a single patch with a semi-infinite superstore with re- lative permittivity equal to unity and the single relative dielectric constant (ε_k) given as:

$$\varepsilon_k = \frac{2\varepsilon_{eff} - 1 + p}{1 + p}$$

where

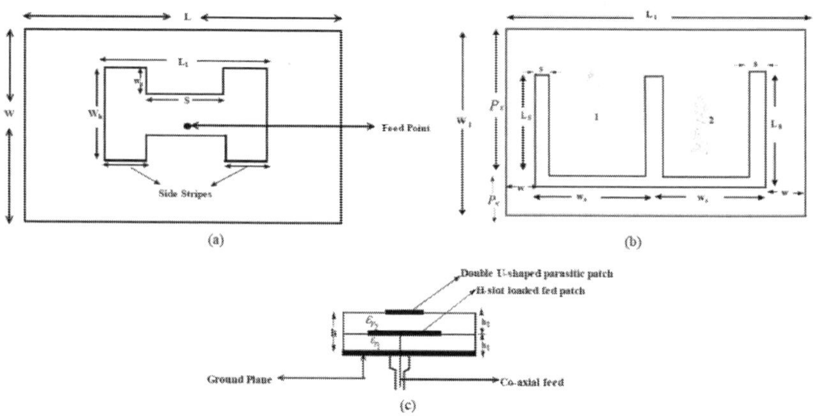

Figure 1. Configuration of stacked patch antenna. (a) H-slot loaded fed patch; (b) Double U-shaped slots loaded parasitic patch; (c) Side view of the proposed antenna.

$$p = 1 + \frac{10h_1}{W_e}$$

in which W_e is the effective width and ε_{eff} is the effective dielectric constant of the structure [13] . The effective dielectric constant of the lower patch is given as

$$\varepsilon_{e_1} = \frac{\varepsilon_k + 1}{2} + \frac{\varepsilon_k + 1}{2}\left(1 + \frac{12h_1}{W}\right) \tag{1}$$

where,

h_1 = height between ground plane and lower patch;

W = width of the patch.

The equivalent circuit of the simple patch antenna is parallel combination of resistance (R_1), inductance (L_1) and capacitance (C_1) (Figure 2), whose values are defined as [14]

$$C_1 = \frac{\varepsilon_0\,\varepsilon_{e_1}\,L\,W}{2h_1}\cos^{-2}\left(\frac{\pi y_0}{L}\right) \tag{2}$$

$$L_1 = \frac{1}{\omega_1^2 C_1} \tag{3}$$

$$R_1 = \frac{Q_1}{\omega_1 C_1} \tag{4}$$

where

$$Q_1 = \frac{c\sqrt{\varepsilon_{e_1}}}{4 f_1 h_1}$$

L = length of the lower patch;

y_o = Y-coordinate of the feed point.

$$\omega_1 = 2\pi f_1$$

$$f_1 = \frac{c}{2(L + 2\Delta L)\sqrt{\varepsilon_{eff}}}$$

c = velocity of light.

ΔL = fringing length for the lower patch.

Considering the top patch as a simple stacked rectangular microstrip patch, the values of resistance (R_2), inductance (L_2) and capacitance (C_2) can be given as

$$C_2 = \frac{\varepsilon_0\, \varepsilon_{r_2} L_2 W_2}{2h_2} \tag{5}$$

$$L_2 = \frac{1}{\omega_2^2 C_2} \tag{6}$$

$$R_2 = \frac{Q_2}{\omega_2 C_2} \tag{7}$$

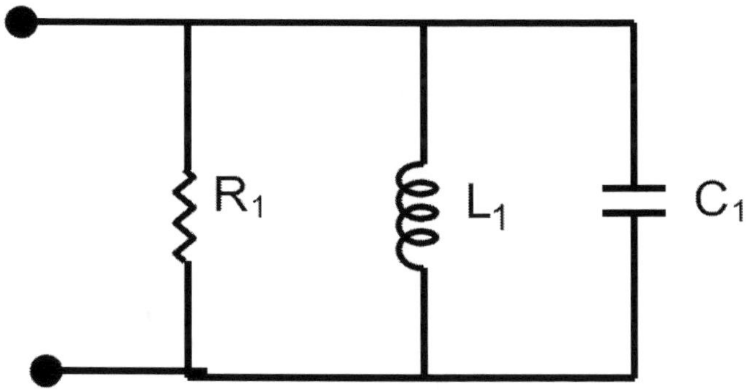

Figure 2. Equivalent circuit of patch antenna.

where

$$Q_2 = \frac{c\sqrt{\varepsilon_{e_2}}}{4 f_2 h_2}$$

where

L_2 = length of the stacked patch;

W_2 = width of the stacked patch;

$$\omega_2 = 2\pi f_2\,;$$

$$f_2 = \frac{c}{2(L_2 + 2\Delta L_2)\sqrt{\varepsilon_{e_2}}}\,;$$

ΔL_2 = fringing length for the top patch.

The equivalent circuit of the fed patch is shown in Figure 3(a), in which ΔL_1 and ΔC_1 are the additional inductance and capacitance respectively, which originate due to introducing the two notches and R_H is resonance resistance after cutting the notches into the patch. The value of R_H can be calculated using Equation (7) and the ΔL_1, ΔC_1 can be given as [15]

$$\Delta L_1 = \frac{h_1 \mu_0 \pi}{8}\left(\frac{W_h}{w_n}\right) \tag{8}$$

where

$$\mu_0 = 4\pi 10^{-7}\ \mathrm{H/m}$$

and

$$\Delta C_1 = \left(\frac{W_h}{w_n}\right) C_s \tag{9}$$

where C_s is the gap capacitance between two side strips [16] . Now the equivalent circuit of H-shaped patch is given as shown in Figure 3(b) in which "Z_N" is the impedance of the notch incorporated patch and is calculated from Figure 3(a), Z_P is the impedance of the initial patch and C_m and L_m are the capacitive and inductive coupling between two resonant circuits.

When a dual U-slots is cut into the stacked patch, current distribution changes which ultimately changes the resonance behavior of the patch. Due to this changing in the patch adds series inductance (ΔL_2) and series capacitance (ΔC_2) in the initial circuit of the stacked patch, which is shown in Figure 4, in which the resonance resistance R_2, ΔL_2 and ΔC_2 are given as [17] - [19]

$$R_2 = \frac{Q_2 h_2}{\pi f_r \varepsilon_{eff} \varepsilon_0 W_2 L_2} \cos^2\left(\left(\left(\frac{\pi y_0}{L_{eff_2}}\right)\right)\right) \tag{10}$$

where ℓ_{eff2} is the effective length [15] and can be given as

$$L_{eff_2} = L_2 + \left(\sin\left(\pi w / L_2\right)\right)\frac{L_s}{2}$$

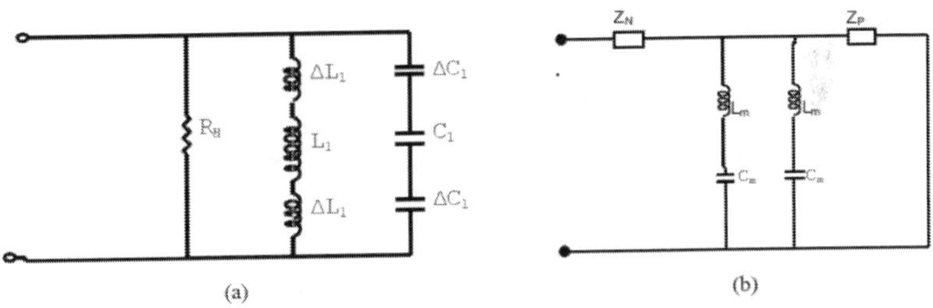

Figure 3. (a) Equivalent circuit of RMSA due to notch effect; (b) Equivalent circuit of H- shaped fed patch.

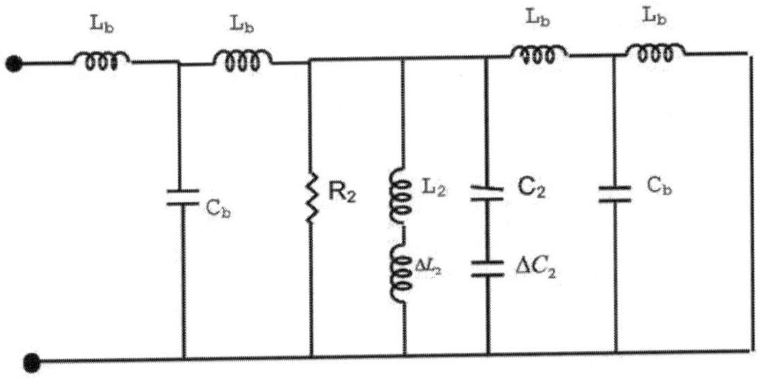

Figure 4. Equivalent circuit of U-slot loaded patch.

$$\Delta L_2 = \frac{Z_1 + Z_2}{16\pi f_1 \cos^{-2}\left(\frac{\pi y_0}{P_E}\right)} \tan\left(\frac{\pi f_1 L_s}{c}\right) \tag{11}$$

where

$$P_E = W_1 - P_N\,;$$

$$Z_1 = \frac{120\pi}{\frac{w_s}{h_2} + 1.393 + 0.667\log\left(\frac{w_s}{h_2} + 1.44\right)};$$

$$Z_2 = \frac{120\pi}{\frac{w_s - 2s}{h_2} + 1.393 + 0.667\log\left(\frac{W_s - 2s}{h_2} + 1.44\right)};$$

ΔC_2 is calculated as gap capacitance and given by [18] .

The value of C_b and L_b are calculated as [19]

$$\frac{C_b}{w} = \left(9.5\varepsilon_{r_2} + 1.25\right)\frac{w}{h_2} + 5.2\varepsilon_{r_2} + 7.0 \text{ PF/m} \quad (12)$$

$$\frac{2L_b}{h_2} = 100\left(4\frac{w}{h_2} - 4.21\right) \text{NH/m} \quad (13)$$

Similarly we can analyze circuit concept of second U-slot which is parallel and compact to first U-slot on the upper patch.

The equivalent circuit of the proposed multiband antenna can be given as shown in Figure 5, in which only capacitive coupling is considered and given as [20]

$$C'_m = \frac{\left(C_{eq} + C'_{eq}\right) + \sqrt{\left(C_{eq} + C'_{eq}\right)^2 - 4C_{eq}C'_{eq}\left(1 - k_c^{-2}\right)}}{2} \quad (14)$$

where

$$C_{eq} = \frac{\Delta C_1 C_p}{2C_p + \Delta C_1}$$

$$C'_{eq} = \frac{\Delta C_2 C_2}{2C_2 + \Delta C_2}$$

and k_c is the coupling coefficient between two resonators.

Thus the total input impedance can be calculated from Figure 5 as

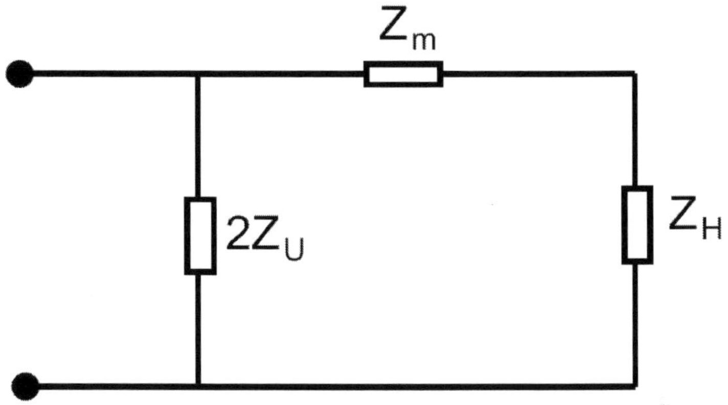

Figure 5. Equivalent circuit of proposed stacked patch antenna.

$$Z_{in} = \frac{2Z_U\left(Z_m + Z_H\right)}{2Z_U + Z_H + Z_m} \tag{15}$$

In which Z_H and Z_U are the impedances of initial and parasitic patches calculated from Figure 3(b) and Figure 4 respectively and Z_m is the impedance due to mutual coupling between fed patch and stacked patch.

3. DESIGN SPECIFICATIONS FOR THE PROPOSED ANTENNA

Substrate material used	FR4
Dielectric constant ($\varepsilon_{r_1}, \varepsilon_{r_2}$)	4.4
Thickness between ground and lower patch (h_1)	1.6 mm
Thickness between lower and parasitic patch (h_2)	1.6 mm
Length of the fed patch (L)	39.25 mm
Width of the fed patch (W)	29.25 mm
Depth of the notch (w_n)	4 mm
Width of the notch (S)	15 mm
Length of the first U slot (L_s)	15.7 mm
Width of the first U slot (s)	1.0 mm
Base width of first U-slot (w_s)	10.3 mm
Length of the second U slot (L_s)	15.7 mm
Width of the second U slot (s)	1.0 mm
Base width of second U-slot (w_s)	10.3 mm
Feed location (x_0, y_0)	(−0.275, −4.37 mm)
Length of the parasitic patch (L_2)	39.25 mm
Width of the parasitic patch (W_2)	29.25 mm

4. DISCUSSION OF RESULTS

All the simulation results of the proposed antenna are given as Figures 6-8. Discussion of results have been explain in terms of S_{11} parameters, resonance frequencies, gains and radiation patterns.

The variation of reflection coefficient with frequency for the proposed multiband antenna is depicted in Figure 6; it is found that the antenna can be operated for seven band applications. In which both lower and upper resonance frequencies of each bands are (4.24 GHz and 4.50 GHz), (5.02 GHz and 5.25 GHz), (7.84 GHz and 8.23 GHz), (12.16 GHz and 12.35 GHz), (14.25 GHz and 14.76 GHz), (15.25 GHz and 15.51 GHz), (17.52 GHz and 17.86 GHz) with return loss −13.81 dB, −17.02 dB, −26.07 dB, −12.79 dB, −20.42 dB, −31.32 dB and −18.41 dB respectively. The bandwidth of each bands are 6%, 4.5%, 4.83%, 2.36%, 3.53%, 1.68% and 1.91%. From Figure 7, it is observed that proposed antenna gains are 2.80 dBi, 4.39 dBi, 4.54 dBi, 6.25 dBi, 10.26 dBi, 8.36 dBi and 9.91 dBi respectively. From both the figure it is clear that the antenna gain is high as a higher frequency of the proposed antenna so that it is very useful in radar application. In [4] , Ansari et al. has already presented the similar work for dual band application. But in the proposed antenna we have replace and changed the radiating patch in place of U-slot loaded patch and H-slot loaded parasitic patch. We have to see that, if changed

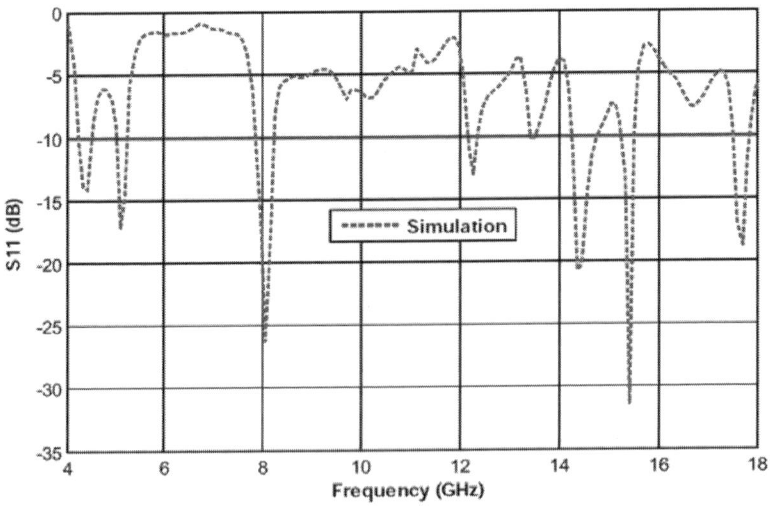

Figure 6. Variation of return loss with frequency for the proposed antenna.

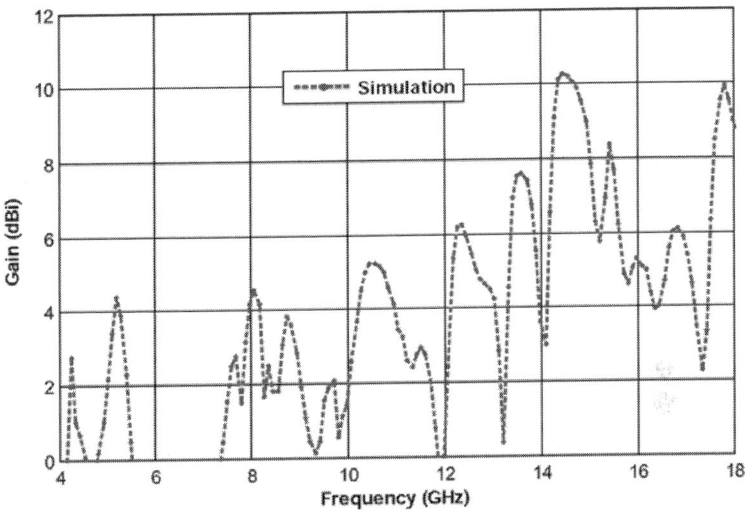

Figure 7. Variation of gain with frequency for the proposed antenna.

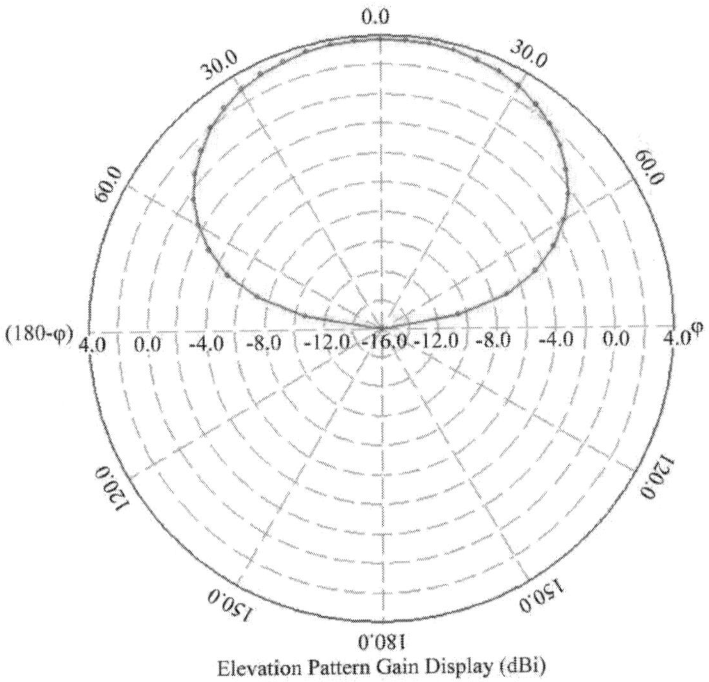

Figure 8. Radiation pattern of the proposed antenna at frequency 8.08 GHz.

the patch, the results are obtained as a multiband antenna with good gain, which is more useful as compared to dual band application presented in 2008. Radiation pattern of the proposed antenna is shown in Figure 8. It is found that radiated power is good. This shows that the directivity improves by stacking dual U-shaped patch.

5. CONCLUSION

A stacked multiband proximity coupled microstrip patch antenna is presented. This antenna has a very simple structure printed on FR4 substrate. Multiband has achieved by using dual U-slot loaded stacked patch. The total volume of the antenna is 39.25 × 29.25 × 1.6 mm^2. The proposed antenna shows satisfactory multiband performance and good radiation pattern. Proposed antenna finds the application in C, X and Ku band which can be used for radar, VSAT, mobile and satellite communication.

REFERENCES

1. Lee, K.F., Lung, S., Yang, S. and Kishk, A.A. (2008) Dual- and Multiband U-Slot Patch Antennas. IEEE Antennas and Wireless Propagation Letters, 7, 645-647.
2. Ansari, J.A., Yadav, N., Singh, P. and Mishra, A. (2009) Compact Half U Slot Loaded Shorted Rectangular Patch Antenna for Broadband Operation. Progress in Electromagnetic Research M, 9, 215-226.
3. Ansari, J.A., Mishra, A., Yadav, N., Singh, P. and Vishvakarma, B.R. (2012) Analysis of W-Slot Loaded Patch Antenna for Dual-Band Operation. International Journal of Electronics, 66, 32-38.
4. Ansari, J.A., Singh, P., Dubey, S.K., Khan, R.U. and Vishvakarma, B.R. (2008) H-Shaped Stacked Patch Antenna for Dual Band Operation. Progress in Electromagnetic Research B, 5, 291-302.
5. Dahele, J.S., Lee, K.F. and Wond, D.P. (1987) Dual Frequency Stacked Annular Ring Microstrip Antenna. IEEE Transactions on Antennas and Propagation, 35, 1281-1285.
6. Kumar, G. and Ray, K.P. (2011) Stacked Gap-Coupled Multiresonator Rectangular Microstrip Antennas. IEEE AP-S International Symposium Digest, Boston, 514-517.

7. Motin, M.A., Hasan, M.I., Habib, M.S. and Sheikh, M.R.I. (2013) Design of a Modified Rectangular Patch Antenna for Quad Band Application. International Conference on Informatics, Electronics & Vision (ICIEV), Dhaka, 17-18 May 2013, 1-4.

8. Ryu, H.-C., Ahn, H.-R., Lee, S.-H. and Park, W.S. (2002) Triple-Stacked Microstrip Patch Antenna for Multiband System. Electronics Letters, 38, 1496-1497.

9. Motin, M.A., Hasan, M.I. and Asaduz-Zaman, M. (2012) Design and Optimization of a Low Cost Multi Band Microstrip Patch Antenna for K-Band, Ku-Band and X-Band Application. 15th International Conference on Computer and Information Technology (ICCIT), Chittagong, 22-24 December 2012, 615-620.

10. Li, J.X., Shi, H.Y., Li, H. and Zhang, A.X. (2014) Quad-Band Probe-Fed Stacked Annular Patch Antenna for GNSS Applications. IEEE Antennas and Wireless Propagation Letters, 13, 372-375.

11. Ahsan Ali, S., Rafique, U., Khan, A.U. and Arif, M. (2012) Multiband Microstrip Patch Antenna for Microwave Applications. IOSR Journal of Electronics and Communication Engineering, 3, 43-48.

12. Tulintseff, A.N., Ali, S.M. and Kong, J.A. (1991) Input Impedance of a Prob Fed Stacked Circular Microstrip Antenna. IEEE Transactions on Antennas and Propagation, 39, 381-390.

13. Jiri, S (1992) Analysis of Multilayer Microstrip Lines by a Conformal Mapping Method. IEEE Transactions on Microwave Theory and Techniques, 40, 769-772.

14. Bahal, I.J. and Bhartia, P. (1980) Microstrip Antennas. Artech House, Dedham.

15. Deshmukh, A.A. and Kumar, G. (2007) Formulation of Resonant Frequency for Compact Rectangular Microstrip Antennas. Microwave and Optical Technology Letters, 49, 498-501.

16. Garg, R. and Bahal, I.J. (1978) Microstrip Discountinuities. International Journal of Electronics, 45, 81-87.

17. Zhang, X.X. and Yang, F. (1998) Study of a Slit Cut Microstrip Antenna and Its Application. Microwave and Optical Technology Letters, 18, 297-300. http://dx.doi.org/10.1002/(SICI)1098-2760(199807)18:4<297::AID-MOP14>3.0.CO;2-1

18. Bahal, I.J. (2003) Lumped Elements for RF and Microwave Circuits. Artech House, Boston.

19. Abboud, F., Damiano, J.P. and Papiernik, A. (1988) Simple Model for the Input Impedance of Coax-Fed Rectangular Microstrip Patch Antenna for CAD. IEE Proceedings H Microwaves, Antennas and Propagation, 135, 323-326.

20. Hoefer, J.R.W. (1997) Equivalent Series Inductivity of a Narrow Transverse Slit in Microstrip. IEEE Transactions on Microwave Theory and Techniques, 25, 822-824.

12

CHAPTER

RESONANT LENGTH FORMULATIONS FOR DUAL BAND SLOT CUT EQUILATERAL TRIANGULAR MICROSTRIP ANTENNAS

Amit A. Deshmukh[1], Kamala Prasan Ray[2]

[1]*DJSCOE, Vile – Parle (W), Mumbai, India;*

[2]*SAMEER, Mumbai, India.*

ABSTRACT

The dual band equilateral triangular microstrip antennas are realized by cutting the slots of either quarter wave or half wave in length, inside the patch. In this design, however these simpler approximations of slot length against the frequency do not give closer results for different slot lengths and there positions inside the patch. In this paper, the modal variations of slot cut patch antennas over wide frequency range are studied. It is observed that the slot does not introduce any mode but reduces the higher order mode resonance frequency of the patch and along with the fundamental mode realizes dual band response. The formulations of the resonant length for the mode introduce by the slots in these antennas are proposed. The resonance frequencies calculated using proposed formulations agree well with the simulated results with an error of less than 5%.

Keywords: Equilateral Triangular Microstrip Antenna, Dual Band Microstrip Antenna, U-Slot, Pair of Rectangular Slots

1. INTRODUCTION

The dual band microstrip antenna (MSA) is realized by cutting a slot at an appropriate position inside the patch [1-3]. Since the slots are cut inside the patch, they neither increase the patch size nor largely affect the radiation pattern of the patch. These slots can take different shapes like, rectangular or square slot, step slot, tooth-brush shaped slot, V-slot, U-slot, etc [4-8]. The slot adds another resonant mode near the fundamental mode of the patch and realizes dual frequency response. The dual band equilateral triangular MSA (ETMSA) have also been realized by cutting the pair of rectangular or U-slot inside the patch [4-9]. While designing these slotted MSAs for the required frequency, depending upon where they are cut in the patch, the slot length is taken equal to nearly half wave or quarter wave in length. Towards the shorted end of the slot, the surface currents encircles over a finite length, which stores magnetic energy towards the shorted slot end [10]. To account for this an additional length is added to this half or quarter wavelength. This additional length is reported to be 10% of the slot length at the required frequency [10]. However it was observed that these simpler approximations of slot length against the frequency do not give closer results for varying slot lengths and their positions inside the patch. Therefore an in-depth analysis for the slot cut MSA is required over a wide frequency range so as to accurately model the slot length at the required frequency.

In this paper, the slot frequencies for U-slot or rectangular slots cut dual band ETMSAs, calculated by using their half wave or quarter wave length approximations are compared with the simulated values obtained using IE3D software [11]. It was observed that the above approximations give larger error for wide range of slot lengths and their positions inside the patch. Therefore a detail modal analysis for these slots cut ETMSAs was carried out over a wide frequency range using the surface current distributions obtained using IE3D software. It was observed that the U and rectangular slots alters the resonance frequency of higher order TM_{11} mode of ETMSA and along with its fundamental TM_{10} mode realizes dual band response. The radiation pattern at the TM_{10} mode for ETMSA is in the broadside direction whereas at TM_{11} mode it is in the conical direction. Since the slot affects the current distribution on the patch at TM_{11} mode, with the increasing slot length the radiation pattern at the modified TM_{11} frequency becomes in the broadside direction. The cross-polarization levels decreases with the increase in slot length. The surface current distributions for slot cut ETMSAs were studied for different slot lengths and the resonant length formulation for the slot (modified higher order) mode and patch mode is proposed. The resonant frequencies calculated using the proposed formulations agrees well with the simulated results with an error of less than 5% over the entire slot length range. The formulations are first proposed on glass epoxy substrate (e_r = 4.3, h = 0.16 cm, and tan d = 0.02) and further they are validated on RT-Duroid substrate (e_r = 2.33, h = 0.16 cm, and tan d = 0.001).

2. DUAL BAND SLOT CUT ETMSAS

In most of the reported slotted antennas the slot length is taken to be equal to half or quarter wave in length. In dual band pair of rectangular slots cut ETMSA as shown lengths (l) from 0.5 to 4.0 cm, the equivalent slot resonant length is calculated by adding the correction length as given in Equation (1). The slot frequency is calculated by using Equation (2). Since the surface currents circulate around the slot, the effective dielectric constant (e_{re}) is calculated by using the average of the widths on the two sides of the slot [1]. For different values of Y, the frequencies calculated using Equations (1) and (2) are plotted against the simulated frequencies obtained using IE3D in Figure 1(b-d). The % error (E) with respect to the simulated value is calculated by using Equation (3) and it is also plotted in the same figure.

$$l_e = l + \frac{0.4l}{\sqrt{\varepsilon_{re}}} \tag{1}$$

$$f_r = \frac{c}{4l_e\sqrt{\varepsilon_{re}}} \tag{2}$$

$$E = \frac{f_{ie3d} - f_r}{f_{ie3d}} \times 100 \tag{3}$$

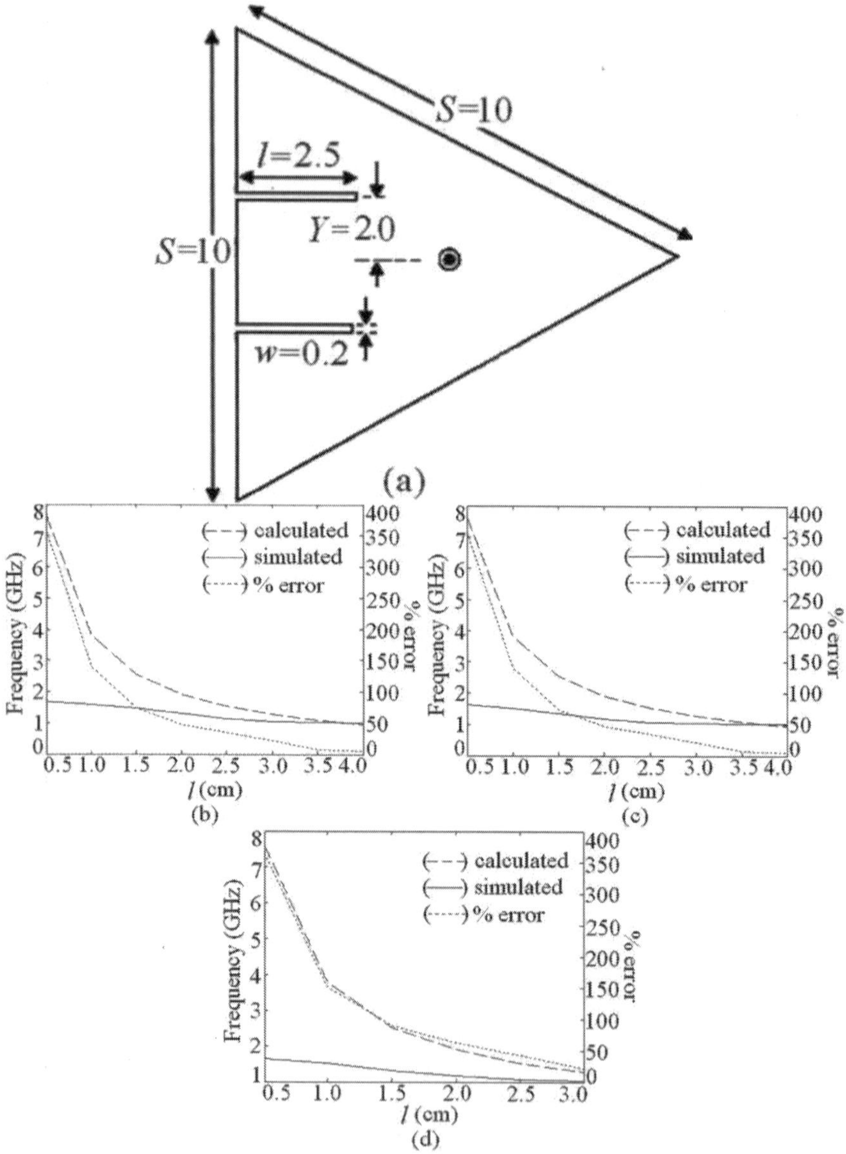

Figure 1. (a) ETMSA with pair of slots and its resonance frequency and % error plots for (b) Y = 1; (c) Y = 2 and (d) Y = 3.

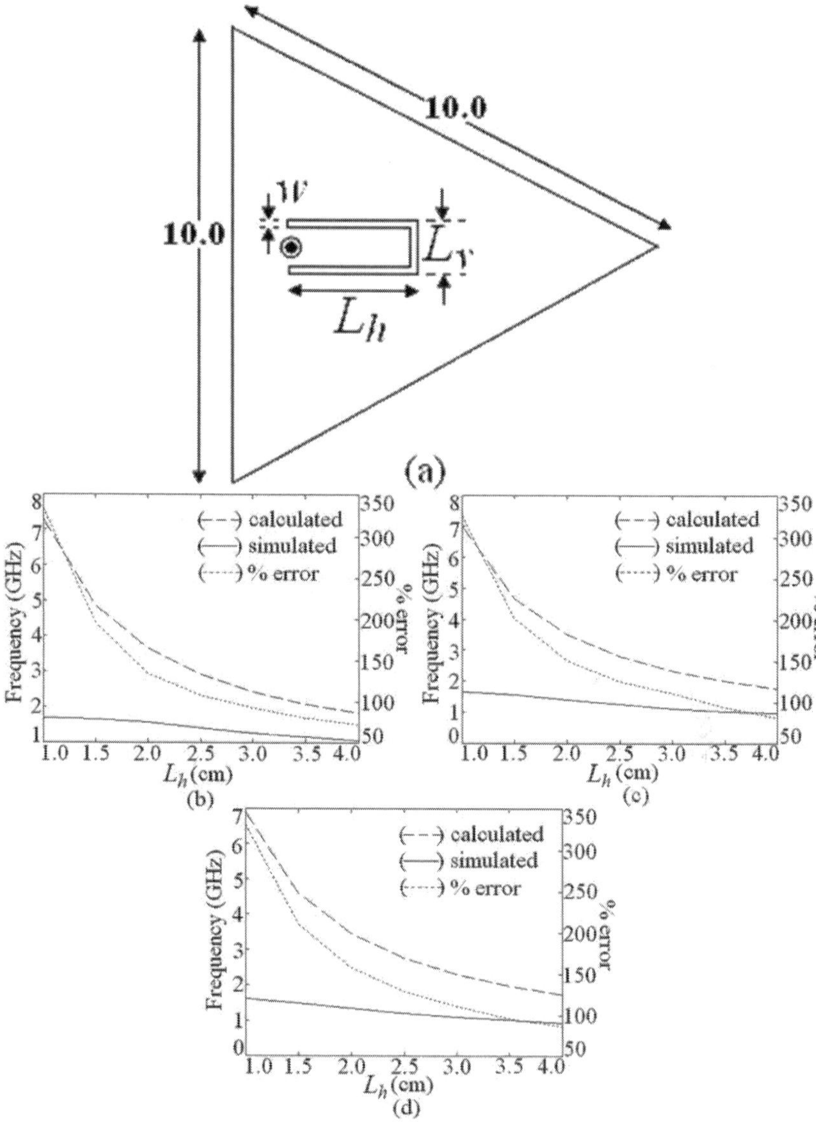

Figure 2. (a) U-slot cut ETMSA and its resonance frequency and % error plots for (b) $L_v = 1$; (c) $L_v = 2$ and (d) $L_v = 3$.

Using this approach for some values of Y and only for larger l, a smaller E is obtained. The dual band U-slot cut ETMSA is shown in Figure 2(a). The inner U-slot length equals half the wave length. The e_{re} is calculated by using the inner U-slot width (L_v – 2w). The resonant length and frequency is calculated by using Equations (4) to (6). For different values of L_v and L_h, the slot frequency calculated by using the following equations and that obtained using IE3D are shown in Figures 2(b-d). For all the values of U-slot dimension, a larger E as calculated by using Equation (3), is observed.

$$l_u = 2(L_h - w) + L_v \tag{4}$$

$$l_e = l_u + \frac{0.2l}{\sqrt{\varepsilon_{re}}} \tag{5}$$

$$f_r = \frac{c}{2l_e\sqrt{\varepsilon_{re}}} \tag{6}$$

Thus the quarter wave or half wave length approximations of the slot length against the frequency do not give closer results. Hence in-depth analysis of slot cut MSAs is required to understand the slot mode. The ETMSA in above configurations, on glass epoxy substrate has, TM_{10} and TM_{11} mode resonance frequencies of 958 and 1690 MHz as shown in its resonance curve plot in Figure 3(a). The surface current distributions at these two modes are shown in Figures 3(b, c). At TM_{10} mode the currents show half wavelength variation along side length (S) whereas at TM_{11} mode the currents show half wave length variations along S as well as along base of the ETMSA [1]. In the dual band ETMSA as shown in Figure 1(a), the slot length l is increased in steps of 0.5 cm from 0.5 to 4.0 cm and the surface current distributions for each value of l and for different slot position (Y) were studied. For slot dimensions of l = 2.5 cm, w = 0.2 cm and Y = 2.0 cm, the dual frequencies are f_1 = 940 MHz and f_2 = 1176 MHz, as shown in the resonance curve plot in Figure 3(a). The current distributions at these two frequencies are shown in Figures 3(d, e). As seen from the current distributions for ETMSA and ETMSA with pair of slots and the resonance curve plot that the pair of rectangular slots perturbs the surface currents and reduces the TM_{11} mode resonance frequency from 1690 to 1176 MHz as shown in Figure 3(a). Here the slots do not introduce any mode but reduces the higher order mode resonance frequency and f_2 is governed by this modified (due to the slots) TM_{11} resonance frequency. The f_1 is governed by TM_{10} mode of ETMSA, as half wavelength variation in surface currents is present along S. This frequency is also marginally affected by slots as shown in Figure 3(a). The radiation pattern for ETMSA at TM_{11}

mode and for slot cut ETMSA for two different slot lengths are shown in Figures 4(a-c). At TM_{11} mode, the pattern is in the conical direction. However since the slots affects the current distributions on the patch, with the increasing slot length the radiation pattern becomes in the broadside direction. Also as observed from the patterns, the cross polarization levels decreases with the slot length.

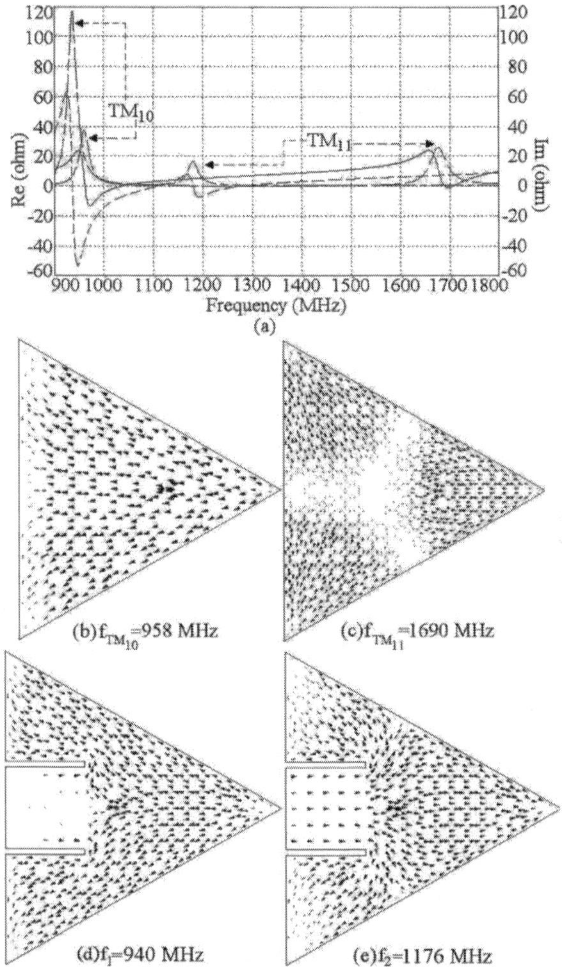

Figure 3. (a) Resonance curve plot for ETMSA (– – –) Re, (—) Im, and ETMSA with pair of slots (— - —) Re, (- - - -) Im, its (b, c) surface current distributions at TM_{10} and TM_{11} modes and surface current distributions for dual slot cut ETMSA at (d) f_1 and (e) f_2.

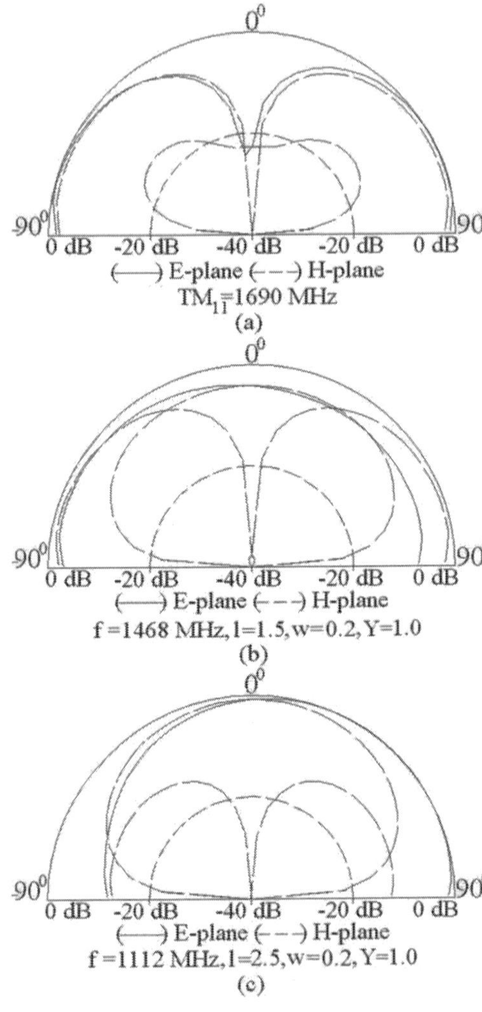

Figure 4. Radiation pattern for ETMSA at (a) TM_{11} mode and at f_2 for (b,c) different lengths of pair of slots.

Similar study is carried out for U-slot cut ETMSA. The resonance curve plot for ETMSA and for U-slot cut ETMSA (U-slot dimension: L_h = 3.0, L_v = 1.0, w = 0.2 cm) are shown in Figure 5(a). The surface current distributions at the dual frequencies for U-slot cut ETMSA is shown in Figures 5(b, c). As seen from the current distributions and resonance curve plots, the horizontal slot of the U-slot reduces the TM_{11} mode resonance frequency and f_2 is governed by this modified TM_{11} mode. This frequency has reduced from

1690 to 1324 MHz. The f_1 is governed by TM_{10} mode. This frequency is also reduced since the vertical U-slot length is orthogonal to surface currents at TM_{10} mode and it increases the surface currents length. The radiation patterns at f_2 for two different U-slot lengths are shown in Figures 5(d, e). Similar to the above dual band ETMSA, the U-slot affects the current distributions on patch at f_2 and gives broadside radiation pattern. The cross polar level decreases with increasing U-slot length. In the following section by studying current distributions at dual frequencies the formulation of resonant length at dual frequencies for the above slot cut ETMSAs are proposed.

3. FORMULATION OF RESONANT LENGTH FOR SLOT CUT ETMSA

The modal variations for ETMSA in voltage and currents are with respect to the centroid of the patch [1]. For S = 10 cm, the centroid point lies at nearly 3 cm from the base of ETMSA. For TM_{10} mode the current maxima is present at the centroid point whereas at TM_{11} mode the voltage maximum is present at the centroid point [1]. Due to these different variations in current and voltage distributions at two modes, for pair of rectangular slots cut ETMSA as shown in Figure 1(a), for l > 3 cm, the decrease in f_2 is smaller whereas decrease in f_1 is larger.

The resonance frequency of ETMSA depends upon the effective side length (S_e) [1]. To formulate f_1 (modified TM_{10} mode) for pair of slots cut ETMSA, S_e is modified with respect to the l and the resonance frequency is calculated by using Equations (7) to (12).

$$W_e = \frac{S}{2} \tag{7}$$

$$\varepsilon_{re} = \frac{\varepsilon_r + 1}{2} + \frac{\varepsilon_r - 1}{2\sqrt{1 + \frac{12h}{W_e}}} \tag{8}$$

$$S_e = S + Al\sin\left(\frac{2\pi Y}{S}\right) + \frac{4h}{\sqrt{\varepsilon_{re}}} \tag{9}$$

$$A = \frac{2.8l}{S}\sqrt{\sin\left(\frac{1.2\pi l}{S}\right)} \tag{10}$$

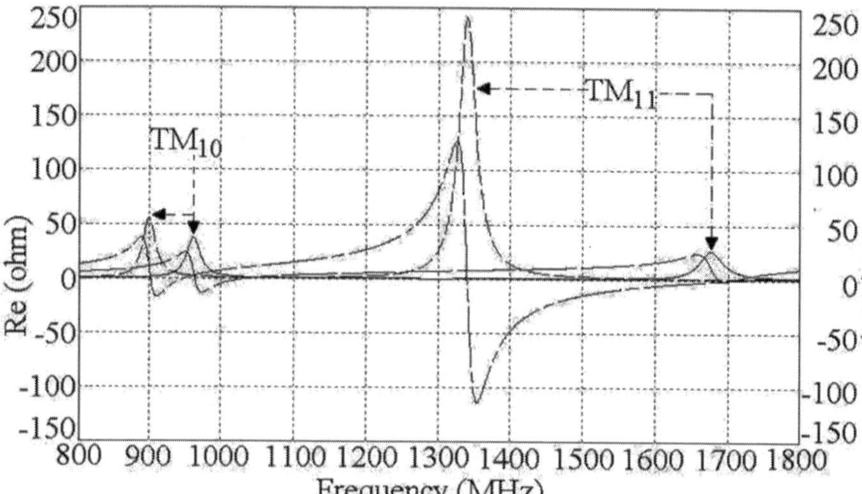

(a)

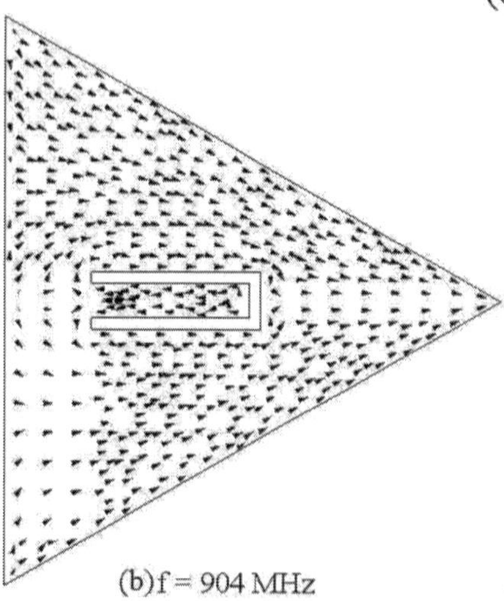

(b) f = 904 MHz

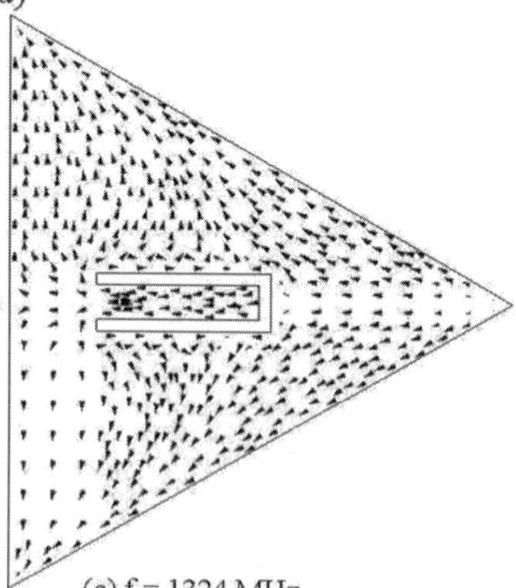

(c) f = 1324 MHz

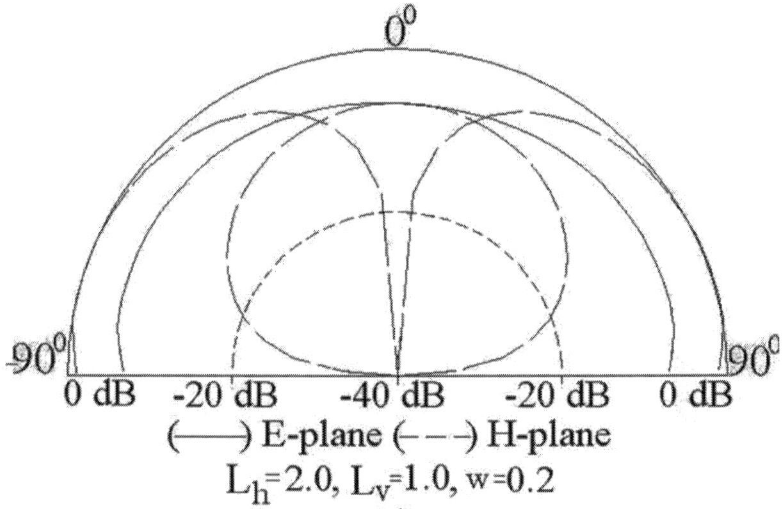

(d)

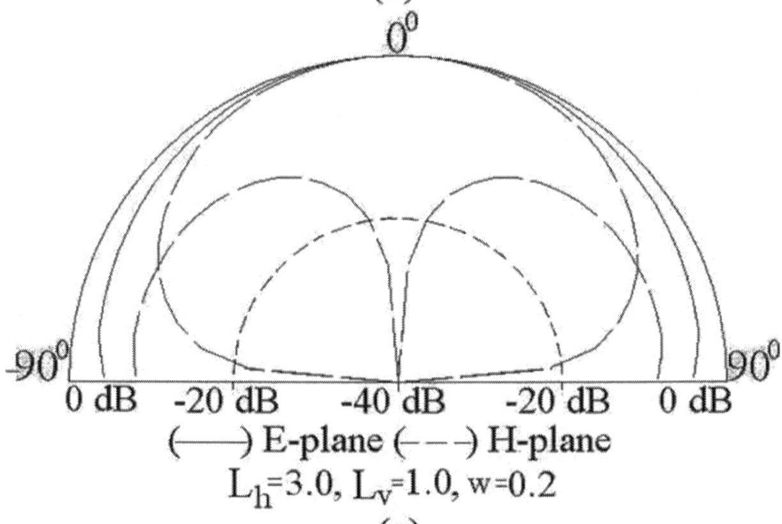

(e)

Figure 5. (a) Resonance curve plot for ETMSA (—) Re, (– – –) Im, and ETMSA with U-slot (----) Re, (—-—) Im, surface current distributions at (d) f_1 and (e) f_2 for U-slot cut ETMSA and its (d,e) radiation pattern for different U-slot lengths.

$$f_1 = \frac{2c\sqrt{m^2+mn+n^2}}{3S_e\sqrt{\varepsilon_{re}}}$$, m = 1, n = 0 for TM_{10} (patch mode) (11)

$$E = \frac{f_{ie3d} - f_1}{f_{ie3d}} \times 100 \quad (12)$$

The width as seen by the surface currents varies from base of ETMSA towards the vertex point. Towards the base, it is equal to S and towards the opposite vertex point it is nearly equal zero. Hence the effective width (W_e) is taken equal to be the average of the width along ETMSA length as given in Equation (7). The e_{re} is calculated by using Equation (8). For smaller l, decrease in f_1 is smaller and for larger l, decrease in f_1 is larger. To account for this non-linear variation, a weighting function (A) in terms of l is used as given in Equation (10). The equation for A is derived based on the variation in frequency of TM_{10} mode against l. For a given l, with increase in Y, the perturbation in surface current length increases (or the frequency reduces) since the slots are placed towards the maximum current location. This variation is accounted by using the sinusoidal term, as given in Equation (9). The fringing field extension is accounted by the last term in the right hand side of Equation (9). The frequency is calculated by using Equation (11) and the % error (E) between the calculated and simulated values with respect to the simulated value, is calculated by using Equation (12) and they are plotted in Figures 6(a-f). For the entire slot length range, an E of less than 5% is obtained. The resonant length at f_2 is formulated by modifying S_e with respect to l as given in Equations (13) to (16).

$$S_e = S + 2Al\sin\left(\frac{2\pi Y}{S}\right) + \frac{4h}{\sqrt{\varepsilon_{re}}} \quad (13)$$

$$A = \frac{5.3l}{S}\sqrt{\sin\left(\frac{2.4\pi l}{S}\right)} \quad (14)$$

$$f_2 = \frac{2c\sqrt{m^2+mn+n^2}}{3S_e\sqrt{\varepsilon_{re}}}$$, m = 1, n = 1 for TM_{11} (slot mode) (15)

$$E = \frac{f_{ie3d} - f_2}{f_{ie3d}} \times 100 \quad (16)$$

The factor '2' in Equation (13) is to account for the circulation of surface currents around the slot length. The perturbation in surface current length is modeled by using the weighting function A and the equation for the same is derived based on the frequency variation with respect to l at TM_{11} mode. The variation in the frequency with Y, for given l, is modeled by using sinusoidal function in Equation (13). The Y is varied from 1 to 3 cm and the l is varied from 0.5 to 4.0 cm. The frequency f_2 obtained

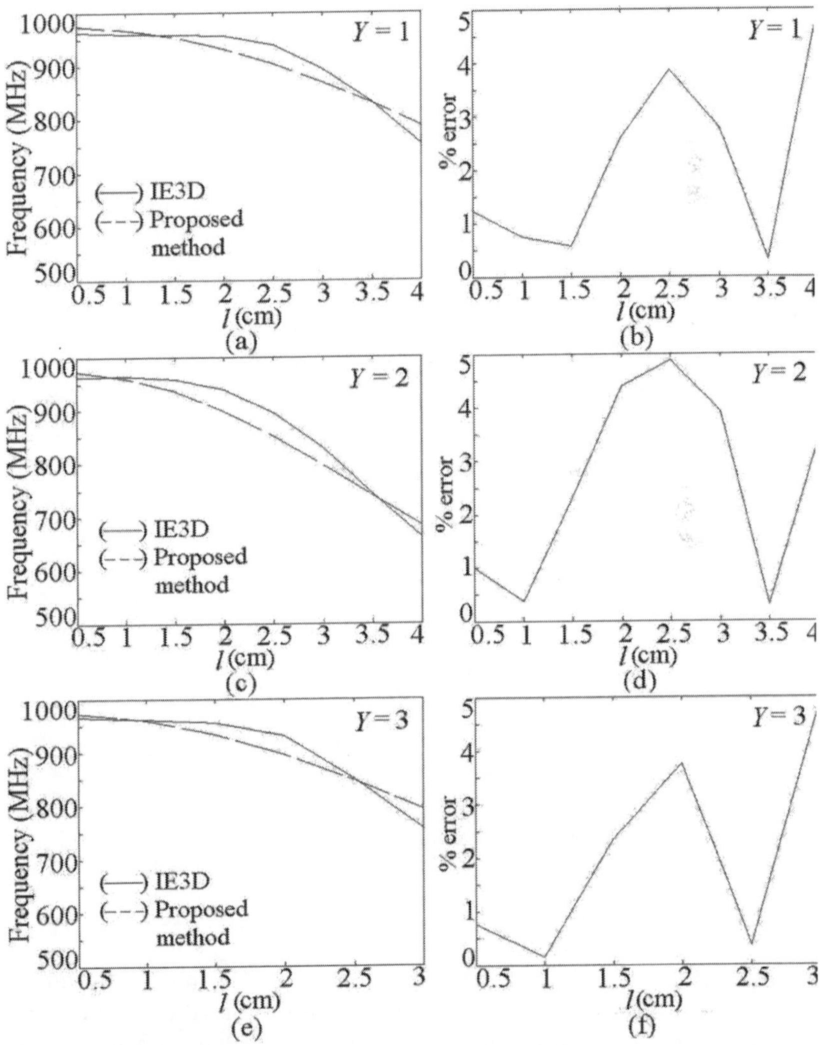

Figure 6. (a-f) Resonance frequency and % error plots for dual band pair of slots cut ETMSA at f_1.

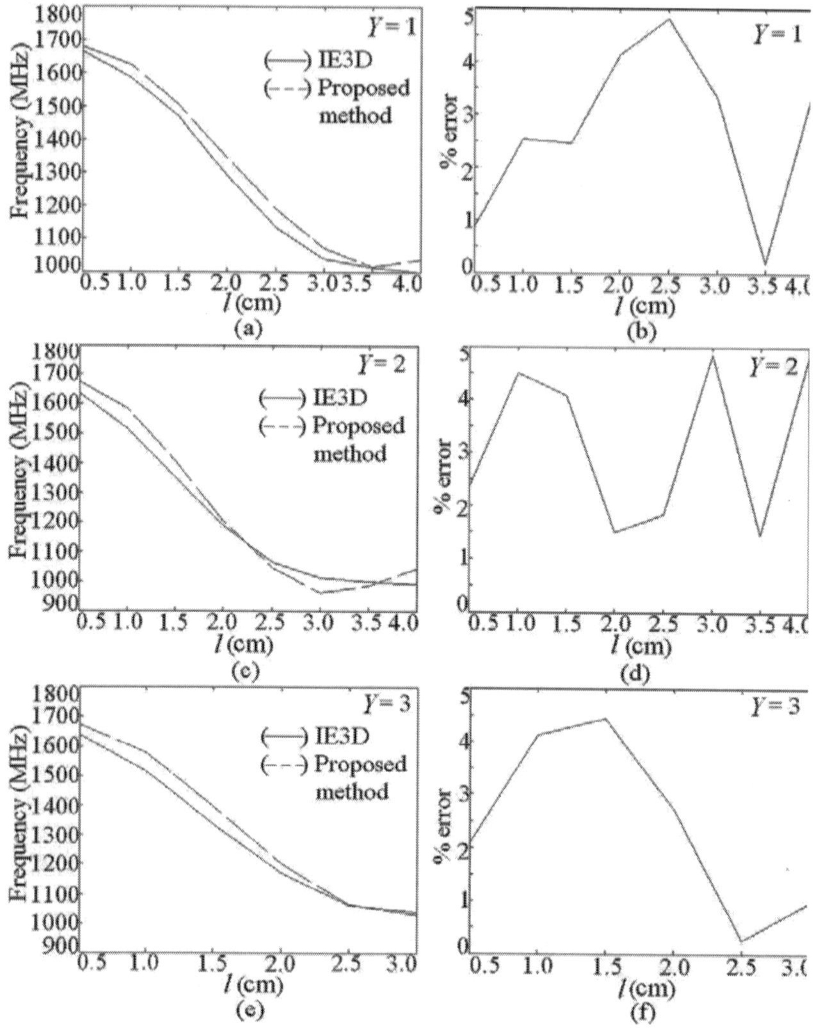

Figure 7. (a-f) Resonance frequency and % error plots for dual band pair of slots cut ETMSA at f_2.

using simulation, calculated by using Equation (15), and E calculated by using Equation (16) are plotted in Figures 7(a-f). For all the values of Y, a close match is obtained between the simulated and calculated frequencies.

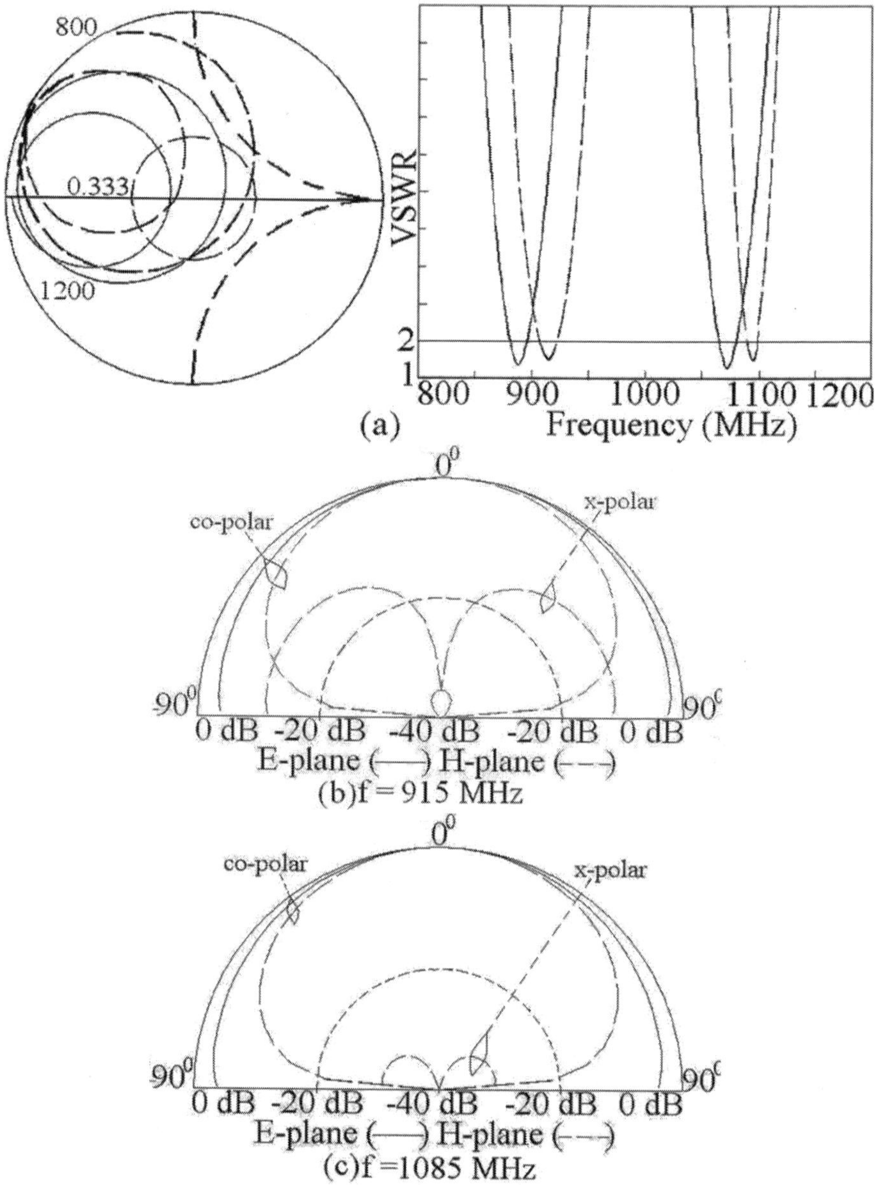

Figure 8. (a) Input impedance and VSWR plots, (—) simulated, (- - -) measured, and (b, c) measured radiation pattern for ETMSA with pair of slots.

While designing the pair of rectangular slots cut ETMSA for the given two frequencies, the separation between the pair of slots is selected first as it affects the impedance matching at the two frequencies. Due to the ETMSA patch geometry, Y is taken to be less than S/4, so as to have larger variation in l and hence the frequency. Thus Y = 2 is selected. For the desired dual frequencies of 860 and 1060 MHz, the slot length obtained from Figures 6 and 7 is = 2.5 cm. The slot width of 0.2 cm is selected. The slotted ETMSA is simulated and dual frequencies and BW's are 888, 1073 MHz and 15, 15 MHz, respectively as shown in Figure 8(a). This response is experimentally verified and the measured frequencies and BW's are 910, 1089 MHz and 12, 13 MHz, respectively as shown in Figure 8(a). The frequency values obtained using proposed equations are very close to the simulated and measured frequencies. The radiation pattern at dual frequencies is shown in Figures 8(b, c). The E and H-planes at both the frequencies are aligned along F = 0^0 and 90^0, respectively. The pattern is in the broadside direction at both the frequencies with the cross polar levels less than 12 dB as compared to the co-polar levels. Although the radiation pattern at TM_{11} mode of ETMSA is conical, but since the pair of slots affects the current distributions on the patch at f_2, the radiation pattern is in the broadside direction.

As seen from the current distributions for U-slot cut ETMSA, the U-slot affects the surface current distributions at TM_{10} as well as TM_{11} mode and realizes dual band response. At f_1, the TM_{10} mode is dominant as half wave length variation in surface currents is present along S. However due to the finite L_v, the surface current length is perturbed and the frequency is reduced. At f_2, the horizontal slots of the U-slot modify the TM_{11} mode resonance frequency and this frequency is governed by modified TM_{11} mode. The formulation of resonant length at f_1 is given by modifying S_e with respect to the U-slot dimensions as given in Equations (17) to (20).

For f_1,

$$W_e = \frac{S}{2} \tag{17}$$

$$\varepsilon_{re} = \frac{\varepsilon_r + 1}{2} + \frac{\varepsilon_r - 1}{2\sqrt{1 + \frac{12h}{W_e}}} \tag{18}$$

$$S_e = S + BL_v + C\frac{L_h}{S} + \frac{4h}{\sqrt{\varepsilon_{re}}} \tag{19}$$

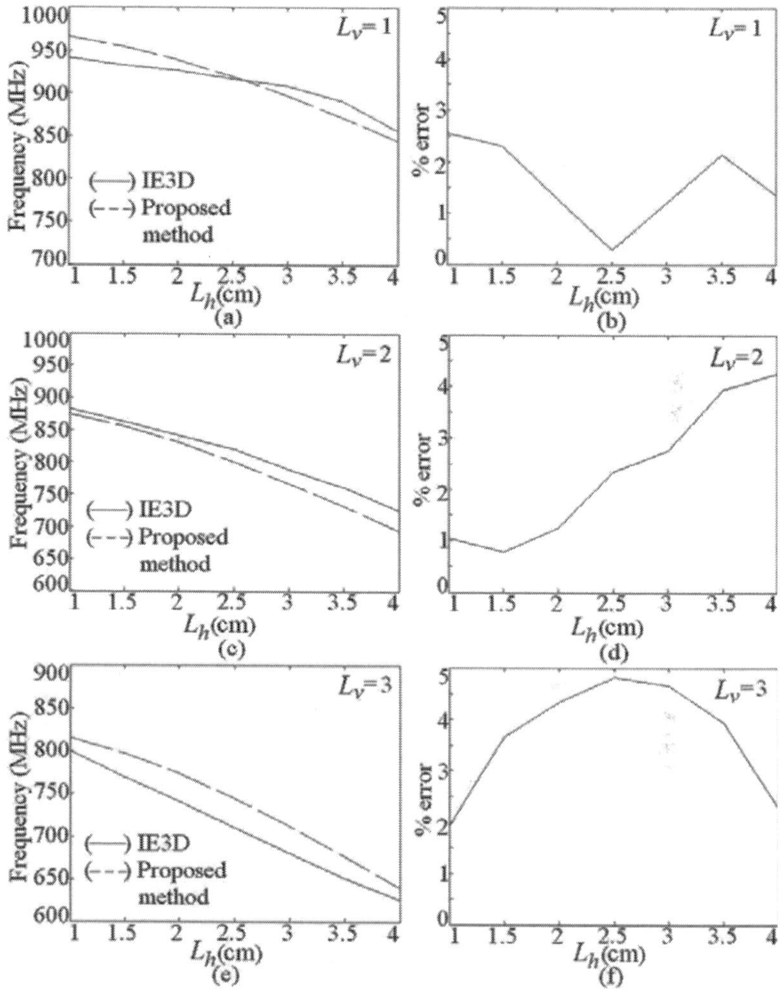

Figure 9. (a-f) Resonance frequency and % error plots for dual band U-slot cut ETMSA at f_1.

$$f_1 = \frac{2c\sqrt{m^2 + mn + n^2}}{3S_e\sqrt{\varepsilon_{re}}}$$, m = 1, n = 0 for TM_{10} mode (20)

For narrow slots, ($L_v < S/4$), the perturbation in surface current length is lesser. In those cases, using B = 0 and A = 1, a closer prediction in resonant length is obtained. For $L_v > S/4$, the perturbation in surface current length increases and hence by using B = 0.6 and A = 2, a closer prediction in the length and the resonance frequency is obtained. The f_1 calculated by using Equation (20) and E calculated by using Equation (12) are plotted in Figures 9(a-f). Using the proposed formulations, for all the slot dimensions, an E of less than 5% is obtained. At f_2, the formulation in resonant length is obtained by using Equations (21) to (23).

For f_2,

$$S_e = S + 2AL_h \sin\left(\frac{2\pi L_v}{S}\right) + \frac{4h}{\sqrt{\varepsilon_{re}}} \tag{21}$$

$$A = \frac{2.8L_h}{S} \tag{22}$$

$$f_2 = \frac{2c\sqrt{m^2 + mn + n^2}}{3S_e\sqrt{\varepsilon_{re}}}$$, m = 1, n = 1 for TM_{11} mode (23)

Similar to pair of slots cut ETMSA, the factor '2' in Equation (21) is to account for the circulation of surface currents around the slot length. To account for the current variation in TM_{11} mode a sinusoidal function in L_v is used in Equation (21). The A is the weighting function and the relation for A is derived based on variation in frequency with respect to U-slot dimensions for TM_{11} mode. The frequency and E are calculated by using Equations (23) and (16), respectively and they are plotted in Figures 10(a-f). For the entire range, frequencies calculated using the proposed formulations agrees closely with the simulated values. For designing the U-slot cut ETMSA, first the value of L_v is selected. For the dual frequencies of 900 and 1300 MHz, using L_v = 1.0 cm and w = 0.2 cm, the horizontal U-slot length obtained by using frequency plots in Figures 9 and 10 is, L_h = 3.0. This ETMSA with U-slot is simulated using IE3D and the frequencies and BW's are 883, 1245 MHz and 15, 18 MHz, respectively as shown in Figure 11. The U-slot cut ETMSA is fabricated and the measured frequencies and BW's are 910, 1275 MHz and 15, 19 MHz, respectively as shown in Figure 11. The frequencies calculated using proposed equations agrees well with the simulated and measured results. The radiation pattern at the two frequencies is in the broadside direction. The E and H-planes are aligned at F = 0^0 and 90^0, respectively. At f_2, a higher cross polarization is present. The cross polar level decreases with the increase in U-slot dimensions. These proposed formulations are also validated on RTduroid substrate. Using duroid substrate, for to have

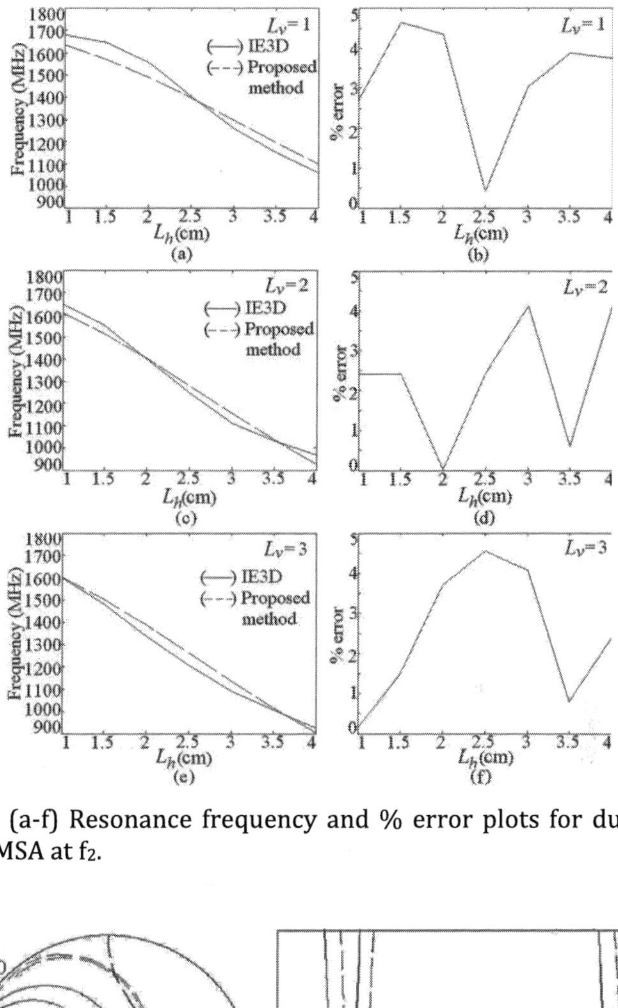

Figure 10. (a-f) Resonance frequency and % error plots for dual band U-slot cut ETMSA at f_2.

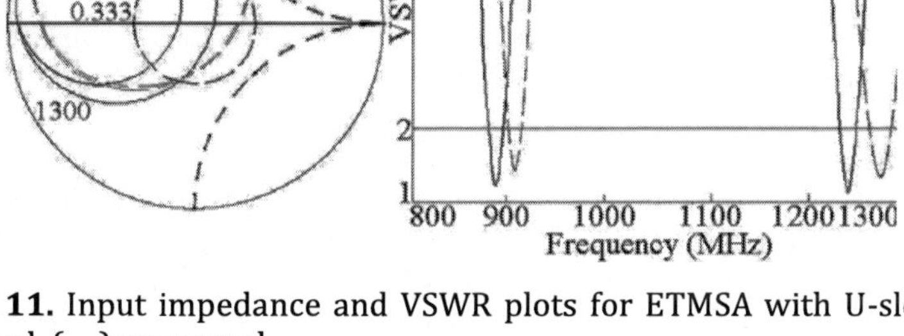

Figure 11. Input impedance and VSWR plots for ETMSA with U-slot, (—) simulated, (---) measured.

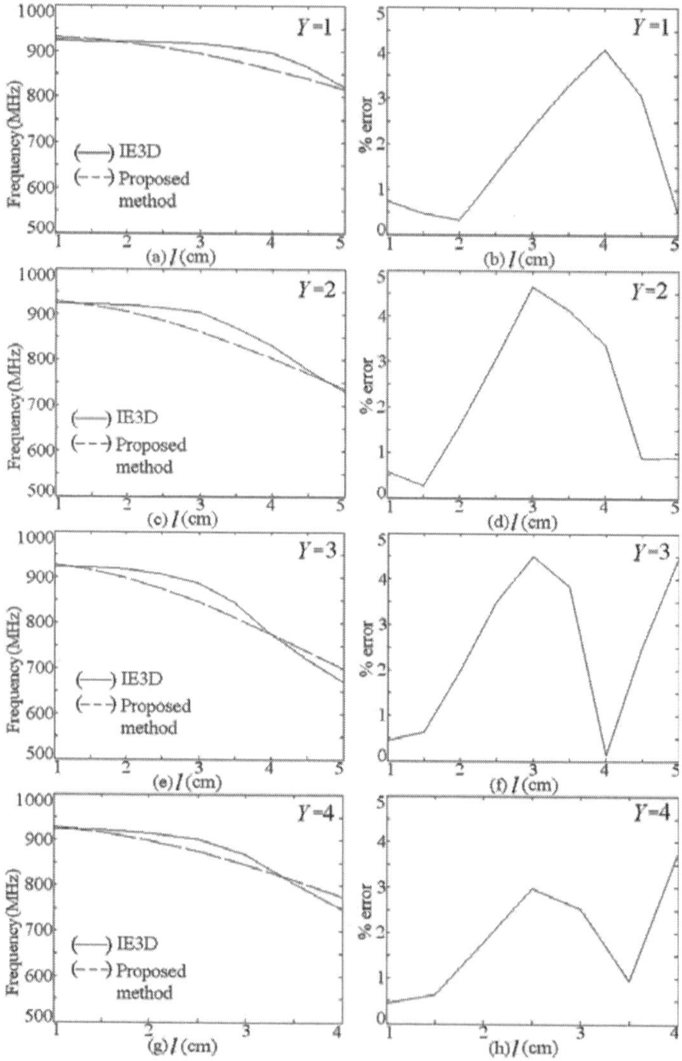

Figure 12. (a-h) Resonance frequency and % error plots for pair of slots cut ETMSA using RT-duroid substrate at f_1.

TM_{10} resonance frequency nearly equal to 950 MHz, the S equals 14 cm. For pair of rectangular slots cut ETMSA, the f_1 and f_2 obtained using IE3D and the proposed formulations and E between them are shown in Figures 12 and 13. For duroid substrate also a closer agreement between the two results is realized. Similar results are obtained for U-slot cut ETMSA.

4. CONCLUSIONS

The resonance frequency calculation for slot mode in pairs of rectangular slots and U-slot cut ETMSAs using quarter and half wavelength approximations are discussed. It was observed that they give closer match only for some values of slot lengths and their positions inside the patch. An in-depth modal analysis for pair of rectangular slots cut and U-slot cut ETMSA was carried out and it is observed that the slots does not introduce any mode but reduces the higher order TM_{11} resonance frequency of ETMSA and along with the fundamental TM_{10} mode realizes dual frequency response. The slots affects the current distribution on the patch at TM_{11} mode and thus with increasing slot length the radiation pattern at the frequency due to the slot becomes in the broadside direction. The cross polarization level reduces with the slot length. By studying the surface current distributions at the dual frequencies, the formulation in resonant length at dual frequencies is proposed for these slotted ETMSAs and the frequencies obtained using these formulations agree closely with the simulated as well as measured results.

REFERENCES

1. G. Kumar and K. P. Ray, "Broadband Microstrip Antenna," Artech House, Norwood, 2003.
2. R. Garg, P. Bhartia, I. Bahl and A. Ittipiboon, "Microstrip Antenna Design Handbook," Artech House, Norwood, 2001.
3. K. P. Ray and D. D. Krishna, "Compact Dual Band Suspended Semicircular Microstrip Antenna with Half U-Slot," Microwave & Optical Technology Letters, Vol. 48, No. 10, 2006, pp. 2021-2024.
4. K. L. Wong, "Compact and Broadband Microstrip Antennas," John Wiley & sons, Inc., New York, 2002.
5. A. E. Daniel and R. K. Shevgaonkar, "Slot-Loaded Rectangular Microstrip Antenna for Tunable Dual-Band Operation," Microwave & Optical Technology Letters, Vol. 44, No. 5, March 2005, pp. 441-444.
6. J. H. Lu, "Single Feed Dual Frequency Rectangular Microstrip Antenna with Pair Step Slots," Electronics Letters, Vol. 35, No. 5, 1999, pp. 354-355.

7. J. Y. Sze and K. L. Wong, "Broadband RMSA Using Tooth-Brush Shaped Slots," Electronics Letters, Vol. 34, No. 23, November 1998, pp. 2186-2187.

8. A. A. Deshmukh and G. Kumar, "Even Mode Multi-Port Network Model for Slotted Dual Band Rectangular Microstrip Antennas," Microwave & Optical Technology Letters, Vol. 48, No. 4, 2006, pp. 798-804.

9. A. A. Deshmukh and K. P. Ray, "Multi-Band Equilateral Triangular Microstrip Antennas," Proceedings of Recent Advances in Microwave Theory and Applications, Jaipur, 21-24 November 2008.

10. J. B. Knorr and J. Saenz, "End Effect in a Shorted Slot," IEEE Transactions on Microwave Theory & Techniques, September 1973, pp. 579-580.

11. IE3D 12.1, Zeland Software, Freemont, 2004.

INDEX